Minidi...
Scie...

PETER MELLETT WITHDRAWN

OXFORD UNIVERSITY PRESS

Oxford University Press, Walton Street, Oxford OX2 6DP

Oxford New York
Athens Auckland Bangkok Bombay
Calcutta Cape Town Dar es Salaam Delhi
Florence Hong Kong Istanbul Karachi
Kuala Lumpur Madras Madrid Melbourne
Mexico City Nairobi Paris Singapore
Taipei Tokyo Toronto
and associated companies in
Berlin Ibadan

Oxford is a trade mark of Oxford University Press

British Library Cataloguing in Publication Data

Data available

Library of Congress Cataloging in Publication Data

Mellett, P. (Peter), 1946– / Peter Mellett.
p. cm.
1. Science—Dictionaries. I. Title.
503—dc20 Q123.M45 1993 92-40594
ISBN 0-19-211680-0

10 9 8 7 6 5 4 3

Printed in Great Britain by
Charles Letts (Scotland) Ltd
Dalkeith, Scotland

A

abdomen In *vertebrates, the part of the body between the *diaphragm and the *pelvis. In *arthropods, a separate body segment that joins to the *thorax. The abdomen contains organs that carry out *digestion, *reproduction, and *excretion.

abiotic factor Each of the non-living parts of the *environment of an organism. Abiotic factors include the climate (*climatic factors) and the structure of the soil (*edaphic factors). In water *habitats, abiotic factors include *pH and *turbidity.

absolute zero A temperature equal to 0 kelvin (0 K), −273° Celsius. It is the coldest temperature possible, because the particles in an object then have the lowest possible *kinetic energy.

AC Abbreviation for alternating current. (see *electric current)

acceleration A measure of how fast the *velocity of an object is changing.

$$\text{Acceleration} = \frac{\text{change in velocity}}{\text{time taken}}.$$

Units of acceleration are metres per second per second (m/s² or ms⁻²). E.g. if a car takes 8 seconds to accelerate from a speed of 16 m/s to 20 m/s, then its acceleration is 0.5 m/s².

accumulator Another name for a *secondary cell, especially the *lead–acid cell.

acetic acid The old-fashioned name for *ethanoic acid.

acetylene The old-fashioned name for *ethyne.

acid A substance that dissolves in water to give *hydroxonium ions (H_3O^+) and so makes a solution with a *pH less than 7. In a solution of a strong acid (e.g. sulphuric acid), nearly all the acid molecules break up to form *ions. Strong acids have a pH less than 2. In a solution of a weak acid (e.g. *ethanoic acid) only some of the molecules form ions. Weak acids have a pH between 2 and 6.9. Acids have a sour taste, turn blue litmus red (*indicator), and donate *protons (H^+ ions) to *bases.

acid rain Rain that is weakly acidic due to dissolved pollution, usually oxides of nitrogen from car exhausts and oxides of sulphur from coal burned in power-stations. Acid rain erodes stone buildings and harms trees and fish and other organisms in rivers and lakes.

acrylic A type of thermoplastic that includes *Perspex and the man-made *fibre Acrilan. (see *plastic)

activation energy An amount of energy that *particles of *reactants must have before they can take part in a *chemical reaction. Heating the reactants increases their total energy. The reaction rate increases because more particles have the required activation energy. A *catalyst lowers the activation energy for a reaction. (see *energy level diagram)

adaptation A change during the *evolution of a type of organism, helping it to live and survive in its *environment. Giraffes have long necks. This helps to adapt them to their environment by enabling them to eat leaves that cannot be reached by other animals.

addition reaction A *chemical reaction where two molecules join together to make one new compound only. E.g. the addition of *bromine to *ethene:

$$H \quad H \qquad\qquad H \quad H$$
$$| \quad | \qquad\qquad | \quad |$$
$$C = C + Br_2 \rightarrow Br - C - C - Br.$$
$$| \quad | \qquad\qquad | \quad |$$
$$H \quad H \qquad\qquad H \quad H$$

adrenal gland An *endocrine gland attached to the top of each kidney. The adrenal glands make a *hormone called adrenalin which prepares the whole body for action. It speeds up heart and breathing rates and stimulates the release of glucose into the blood to supply energy. Adrenalin is made during anger or fright or similar stress.

adrenalin A *hormone. (see *adrenal gland)

aerial (antenna) A piece of wire, a rod, or a dish that is used to transmit or receive radio signals. *Electric currents *oscillate in an aerial connected to a transmitter and produce *radio waves. The reverse process happens at a receiving aerial.

aerobic respiration A type of *respiration that uses oxygen to release energy inside *cells. It changes *glucose from food into carbon dioxide and water. Most organisms use aerobic respiration when making energy for moving and living.

aerofoil An object shaped like an aeroplane wing that experiences an upward thrust when a stream of air moves across it. The upper surface of an aerofoil is more curved than the lower surface. Air has further to travel over the upper surface and so exerts a smaller pressure than the air passing over the lower surface. The upward thrust results from the difference between the two pressures.   ✍

AIDS (acquired immune deficiency syndrome) A disease caused by HIV, the human immunodeficiency virus. HIV destroys a type of white blood cell which makes *antibodies. An AIDS sufferer is open to diseases that uninfected people normally

Aerofoil

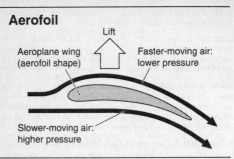

fight off without difficulty. The HIV virus is most commonly passed on during sexual intercourse or when drug misusers share needles and syringes. There is no cure for AIDS at present.

air The mixture of gases that makes up the Earth's *atmosphere.

air pressure see *atmospheric pressure

air resistance The force acting against an object moving through air, caused by *friction between the object and the air around it. The air resistance of a parachute is high; it is low for a *streamlined car. (see *drag)

albinism A lack of pigmentation in an organism. Albino animals and humans have no colour in their skin, eyes, or hair. Albinism results when two *recessive *alleles are inherited.

alcohol 1. A type of *organic compound consisting of a hydroxyl –OH group attached to a *hydrocarbon. The simplest alcohols form a series: methanol CH_3OH; ethanol C_2H_5OH; propanol C_3H_7OH; butanol

C_4H_9OH. They are used in industry as *solvents. 2. The common name for *ethanol.

alga (pl. algae) A type of simple *plant that lives in water or damp places. Brown, green, and red seaweeds are algae. The green slime floating in ponds is an alga called *Spirogyra*.

alimentary canal (gut) In most animals, the long tube through which food passes during *digestion and absorption. In humans, food from the mouth passes down the oesophagus (gullet) and is stored in the *stomach. Digestion starts in the stomach and is completed in the small *intestine. *Villi in the lining of the small intestine absorb digested food into the blood. The large intestine removes water and salts from undigested food. The result is solid waste matter (faeces) which is stored in the (*rectum and passed out through the *anus. (see *peristalsis *duodenum *ileum *caecum *appendix)   ✍

aliphatic Describing an *organic compound that does not contain an *aromatic (benzene) ring of carbon atoms. *Alkanes, *alkenes, *alkynes, and their derivatives are aliphatic.

alkali A *base that dissolves in water to give a solution with a *pH greater than 7. Solutions of alkalis turn red litmus blue (*indicator) and contain *hydroxide (OH^-) ions that react with protons (H^+ ions) from acids. A solution of a strong alkali (e.g. sodium hydroxide) contains high concentrations of hydroxide ions; a solution of a weak alkali (e.g. sodium hydrogencarbonate) contains low concentrations. The pH of a strong alkali is between 12 and 14 and of a weak alkali is between 7.1 and 12.

alkali metal A *metal that is a member of group I in the *periodic table of the *elements. The alkali metals include *lithium, *sodium, and *potassium. They are all extracted by *electrolysis of molten

The human alimentary canal

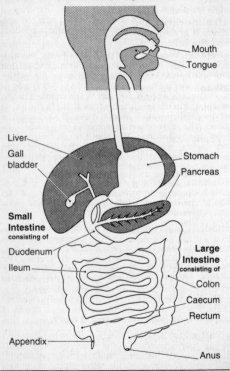

Mouth
Tongue

Liver
Gall bladder

Stomach
Pancreas

Small Intestine
consisting of

Duodenum

Ileum

Large Intestine
consisting of

Colon
Caecum
Rectum

Appendix

Anus

*salts and all are stored under oil. The reactivity of the alkali metals increases down the group. All their compounds are *ionic compounds and their salts are usually soluble in water.

alkaline earth metal A *metal that is a member of group II in the *periodic table of the *elements. The alkaline earth metals include *beryllium, *magnesium, and *calcium. They are all reactive elements extracted by *electrolysis of molten *salts. Their reactivity increases down the group.

alkane A *hydrocarbon molecule that contains carbon–carbon single bonds only and has the general *formula $C_nH_{(2n+2)}$. Alkanes are *saturated and take part in *substitution reactions. They are contained in *crude oil and are used as fuels and to make *organic chemicals.

Name	Formula	Boiling-point (°C)
methane	CH_4	−161
ethane	C_2H_6	−89
propane	C_3H_8	−42
butane	C_4H_{10}	0
pentane	C_5H_{12}	36
hexane	C_6H_{14}	69
heptane	C_7H_{16}	98
octane	C_8H_{18}	126

alkene A *hydrocarbon molecule that contains one or more carbon–carbon double bonds. Alkenes with one double bond have the general *formula C_nH_{2n}. They are *unsaturated and take part in *addition reactions.

alkyne A *hydrocarbon molecule that contains one or more carbon–carbon triple bonds. Alkynes with one triple bond have the general *formula $C_nH_{(2n-2)}$. They are *unsaturated and take part in *addition reactions.

allele Each of the different possible types of a *gene that control a single *characteristic in an organism. E.g. brown and blue eye colour are each controlled by a different gene; the two genes responsible for these two eye colours are called the alleles for eye colour. (see *dominant allele *recessive allele *co-dominance *heterozygous *homozygous)

allergy Some people have an allergy to pollen or dust or some other *antigen. As a result, they suffer from hay fever or asthma. They respond to these antigens by making a substance called histamine which causes swelling and inflammation. Fur and some food and medicines can also cause allergies.

allotropes Two or more forms of an *element with different *structures. *Diamond and *graphite are two allotropes of carbon.

alloy A *substance made by mixing a metal with other metals or non-metals. *Steel (iron+carbon), *brass (copper+zinc), and *bronze (copper+tin) are alloys.

alpha particle A positively charged particle made up from two *protons and two *neutrons. Alpha particles are ejected from the *nuclei of some *radioisotopes. (see *nuclear radiation)

alpha radiation A stream of *alpha particles moving at high speed (up to 1/10 of the speed of light). Its path is bent when travelling close to a magnet or an electrically charged object. Alpha radiation can be stopped by a thick sheet of paper. (see *nuclear radiation)

alternating current (AC) see *electric current

alternative energy Sources of energy that do not depend on *fossil fuels. Alternative energy comes from *nuclear reactors and *renewable energy sources.

alternator An electrical *generator that produces an alternating *electric current.

alum The common name for aluminium potassium sulphate $AlK(SO_4)_2.12H_2O$. Alum is a double *salt whose crystals contain two different metal *ions.

aluminium (symbol Al) A soft silvery-white *metal *element that is a member of group III of the *periodic table. A protective coating of oxide makes it appear less reactive that it really is. Aluminium is extracted by *electrolysis of molten purified *bauxite

Aluminium

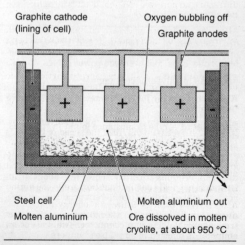

Graphite cathode (lining of cell)

Oxygen bubbling off

Graphite anodes

Steel cell

Molten aluminium

Molten aluminium out

Ore dissolved in molten cryolite, at about 950 °C

Amino acid

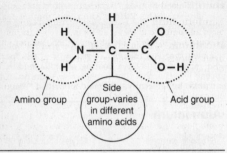

Amino group

Side group-varies in different amino acids

Acid group

---

ore. Because of its lightness and strength, aluminium and its *alloys are used to make aircraft bodies, drinks cans, and overhead power cables. ✍

alveolus (pl. alveoli) see *lung

AM Abbreviation for amplitude modulation. (see *modulation)

amino acid A type of substance found in *proteins; the chemical building-blocks out of which proteins are made. Amino acids contain an amino ($-NH_2$) group and an acid carboxyl ($-COOH$) group. They are made up from carbon, hydrogen, oxygen, and nitrogen. Proteins are broken down into about 20 different amino acids during digestion. Cells use amino acids to make millions of different proteins. ✍

ammeter An instrument used to measure the *electric current flowing in a conductor. The scale of an ammeter is marked in amps. (see *ampere) ✍

ammonia A pungent colourless poisonous gas with the formula NH_3, made by the *Haber process. It is highly soluble in water and forms an *alkaline solution. It is used to make *fertilizers and *nitric acid.

amniotic fluid The liquid that surrounds and protects an *embryo growing in its mother's *uterus. It is contained inside a *membrane called the amnion.

amorphous Describing a solid that is not crystalline e.g. *glass.

ampere (amp; symbol A) The *SI unit of *electric current. 1 A flows through a conductor which has a *resistance of 1 *ohm and with a voltage of 1 *volt applied across its ends. 1 A is equal to 6.24×10^{18} electrons flowing around a circuit each second. A current of about 0.25 A lights the bulb in a table lamp. (see *Ohm's law)

amphibian A type of animal that can live in water or on land but which returns to the water to breed. Amphibian eggs are laid and fertilized in water.

Ammeter

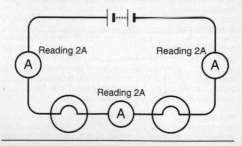

They hatch into tadpoles, which develop by *metamorphosis into the adult form. Amphibians are *vertebrates with four limbs and moist skins and include frogs, newts, and toads.

amphoteric see *oxide

amplifier In *electronics, a circuit that changes a small input signal (a rapidly changing *voltage) into a larger output signal. E.g. an amplifier can change a small input voltage from a *microphone into a larger output voltage across a *loudspeaker. Amplifiers are made up from circuits containing *transistors, *integrated circuits, or *thermionic valves.

amplitude The maximimum size of a *wave or vibration as it *oscillates.

a.m.u. Abbreviation for *atomic mass unit.

amylase A type of *enzyme. Amylases help to break down *carbohydrates in food during *digestion. Human saliva contains an amylase called ptyalin. Digestive juices made in the *pancreas also contain amylases.

anaerobic respiration A type of *respiration that releases energy from food inside living *cells without the use of oxygen. Anaerobic respiration does not produce as much energy as *aerobic respiration.

Amplitude

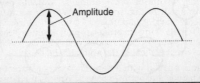

*Yeasts and many types of *bacteria use anaerobic respiration during *fermentation. It also takes place in the muscle cells of animals when there is a shortage of oxygen. (see *oxygen debt)

anaesthetic A local anaesthetic causes loss of feeling in one place. A general anaesthetic causes loss of consciousness. Anaesthetics are used during dental and surgical operations.

analogue A measuring instrument that models the change it is indicating. E.g. a watch with hands is an analogue model of the Earth. The hour hand matches the movement of the Earth. Analogue devices are non-*digital.

angiosperm A *plant that produces seeds with the help of *flowers. Examples include grasses, roses, water-lilies, and *deciduous trees such as oaks and chestnuts.

angle of incidence/reflection see *reflection

anhydrous Describing a substance that contains no water in its *structure. Anhydrous substances are often made by heating *hydrated substances. Water is driven off as steam. E.g.:

blue hydrated copper sulphate crystals		white anhydrous copper sulphate powder	
	heat		steam

$$CuSO_4.5H_2O \longrightarrow CuSO_4 + H_2O.$$

animal A living organism that obtains its food by eating *plants or other animals. Animals usually move about to find their food. *Vertebrate animals include fish, *amphibians, *reptiles, birds, and *mammals. *Invertebrate animals include *coelenterates, *worms, *arthropods, and *molluscs. All animals are members of the animal *kingdom.

anion An *ion with a negative *charge, attracted to the *anode during *electrolysis. Anions include non-metal ions (e.g. oxide O^{2-} and bromide Br^-) and the molecular ions that result when acids dissolve in water (e.g. sulphate SO_4^{2-} from sulphuric acid H_2SO_4).

annelid A type of *worm with a soft cylindrical body divided into segments. Earthworms, lugworms, and leeches are annelids.

annual A type of *herbaceous plant that lives for one year or less. During this time, it grows from a seed, produces flowers and seed, and then dies. Poppies and sunflowers are examples of annual plants. (see *biennial *perennial)

anode An *electrode that has a positive *charge.

anodizing Using *electrolysis to form a thick protective layer of aluminium oxide on the surface of aluminium. The aluminium is the anode and the *electrolyte is dilute sulphuric acid. The oxide layer can be dyed with bright colours.

antagonistic pair Voluntary *muscles are attached to movable *joints in pairs. One muscle (the flexor muscle) bends the joint, the other (the extensor muscle) straightens it. The muscles work against each other and so are called an antagonistic pair.

antenna (pl. antennae) 1. A stalk-like jointed *sense organ found in pairs on the heads of *arthropods. Antennae are responsible for smell, taste, and touch. They may be plain (e.g. ants) or feathery (e.g. some moths). 2. see *aerial

anther see *flower

antibiotic A substance that kills or stops the growth of some *bacteria that cause diseases. Antibiotics are usually injected or swallowed and circulate in the blood and *tissue fluid. Many antibiotics (e.g. *penicillin) are substances made by moulds.

antibody A substance made by *lymphocytes (a type of white blood cell). Antibodies are made when a foreign substance (*antigen) invades the tissues of an animal. Antibodies join with the antigen and make it harmless. Antibodies also help *phagocytes to attack antigens. Each type of antibody has a shape that matches a particular type of antigen. (see *immunity)

anticline see *fold

anticyclone An area of high-pressure air covering part of the Earth. The pressure is highest at the centre. Winds circle around a cyclone and away from the centre. The weather is usually clear and settled.

antigen A foreign substance invading the tissues of an animal. Antigens include *proteins on the outsides of *microbes and *toxins made by microbes. Animals react to antigens by making *antibodies which attack the antigens. This process is called the immune response.

antiseptic A substance that kills or stops the growth of disease-causing *microbes, without damage to human skin *tissue. Antiseptics are used to treat small wounds. Common examples are salt solution and *ethanol.

anus The opening at the end of the *alimentary canal in mammals. Solid waste matter (faeces) passes out through the anus. (see *rectum)

aorta see *heart

apparatus All the sorts of equipment that can be used in *experiments. Examples of common apparatus include containers (e.g. beakers, flasks, plastic bags) and measuring instruments (e.g. rulers, burettes, potometers).

appendicular skeleton see *skeleton

appendix In mammals, a part of the *alimentary canal. It is a closed tube attached to the *caecum. The human appendix plays no part in the *digestion and absorption of food.

appliance A *device or several devices fitted together to make a piece of equipment that performs a particular task. A can-opener is a mechanical appliance, a refrigerator is an electrical appliance, and a radio is an electronic appliance.

aqueous Describing a solution in which the *solvent is water.

aqueous humour see *eye

arachnid A type of *arthropod animal. Arachnids include spiders, scorpions, ticks, and mites. Their bodies are divided into two main parts. The front part carries the sense organs and four pairs of legs. The rear part is the *abdomen, which contains the organs of digestion and reproduction.

arene A *hydrocarbon that contains a *benzene ring of carbon atoms. Arenes undergo *substitution reactions, even though they are are not *saturated compounds.

argon (symbol Ar) A gaseous *element belonging to group VIII of the *periodic table (the *noble gases). Argon is a monatomic gas, composed of separate atoms. It is extracted by *fractional distillation of *liquid air, which contains 0.9% of the gas. Argon is used to fill electric *light bulbs and to protect metals from *oxidation during electric *welding.

aromatic Describing an *organic compound that contains a *benzene ring of carbon atoms. (see *arene)

artery A *blood vessel that carries blood away from the *heart. Arteries have thick muscular walls. All arteries except the pulmonary artery contain oxygenated blood.

arthropod A type of *invertebrate animal. Arthropods are the largest group in the animal *kingdom. They have a hard outer covering (exoskeleton), a segmented body, and jointed legs. They include *crustaceans, *insects, *arachnids, and *myriapods.

artificial satellite A man-made object that moves in an orbit around the Earth, the Moon, the Sun, or a planet. Some satellites take pictures of the world below and others gather information from space. Communications satellites relay information, telephone, TV, and radio signals from one part of the Earth to another. (see *geostationary satellite)

asbestos A *mineral that exists as *fibres that are resistant to heat and chemicals. It is used in vehicle brakes and to make fireproof cloth. Asbestos fibres can cause lung diseases if breathed into the lungs.

aseptic Describing equipment or a place that is free from *pathogens and other harmful *microbes. A hospital operating theatre is aseptic.

asexual reproduction A type of *reproduction where young are produced by one single parent organism alone. Asexual reproduction includes *binary fission, *spore formation, and *vegetative reproduction. Some simple animals (e.g. *Hydra*) reproduce asexually by forming a bud. This grows into a young animal which breaks away from the parent.

asteroid A small body in the *solar system, usually moving around the Sun between the *orbits of Mars and Jupiter. Asteroids are rocky and have irregular shapes. The largest has a diameter of about 1000 km and the smallest less than 1 km.

astronomical unit (AU) A unit of distance. 1 AU is the mean distance between the Sun and the Earth, 149 597 870 km.

atmosphere The gases around a body (e.g. planet, star) or inside a closed space (e.g. space capsule). The Earth's atmosphere is air made up from *nitrogen (78%), *oxygen (21%), argon (0.9%), and carbon dioxide (0.03%). The remainder is mostly water vapour and *noble gases.

atmospheric pressure The *pressure of the Earth's atmosphere, caused by the *weight of the air above the place where it is measured. Atmospheric pressure is about 101 kilopascals (kPa) at sea-level. It decreases with increasing altitude. (see *barometer)

atom The smallest part of an *element, made up from *electrons orbiting around a central *nucleus. The nucleus in all atoms contains *protons and *neutrons, except hydrogen which contains 1 proton only. A *neutral atom contains equal numbers of electrons and protons and so has no overall electric *charge. (see *shell)

Atom

A Helium atom

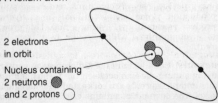

2 electrons
in orbit

Nucleus containing
2 neutrons
and 2 protons ◯

atomic mass unit (a.m.u.) A unit used to measure the mass of a single atom or other small particles. It is equal to 1/12 of the mass of an atom of the isotope carbon-12. 1 a.m.u = $1.660\ 33 \times 10^{-27}$ kg. (see *relative atomic mass)

atomic number (symbol Z) The number of *protons in the nucleus of an *atom. All the atoms in an element have the same atomic number. The atomic number is also the number of *electrons in a neutral (non-ionized) atom.

atrium (auricle) A chamber in the upper part of a *heart. In mammals, the right atrium receives blood from the body and pumps it into the right *ventricle. The left atrium receives blood from the lungs and pumps it into the left ventricle. The two atria have thinner walls than the ventricles.

AU Abbreviation for *astronomical unit.

auditory nerve see *ear

auricle see *atrium

auxin A substance made by cells in the tips of plant stems and roots. Auxin speeds up growth in stems and slows down growth in roots. This controls the response of a plant to gravity and light. (see *tropism)

Avogadro constant (Symbol L or N_A) The number of atoms in 12.000 grams of carbon-12, equal to 6.02×10^{23}. The Avogadro constant is also called the Avogadro number. (see *mole)

Avogadro's law (Avogadro's hypothesis) A law that states: 'Equal volumes of different gases contain the same number of molecules.' E.g. at room temperature and pressure, 24 dm^3 of carbon dioxide and 24 dm^3 of hydrogen each contains 6.02×10^{23} molecules. (see *molar gas volume)

Axle

Well

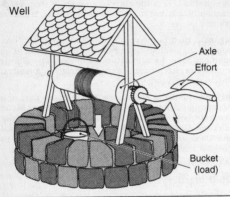

Axle

Effort

Bucket (load)

axial skeleton see *skeleton

axis The line around which an object can rotate or about which it is symmetrical. A bicycle wheel rotates about an axis that passes through its centre.

axle A thin rod (shaft) attached to the centre (*axis) of a wheel. A wheel and axle together make a simple *machine. A small effort at the wheel can move a large load attached to the axle.

axon The long threadlike part of a *neurone (nerve cell).

B

background radiation Small amounts of *nuclear radiation on the Earth and in the atmosphere. It is mainly caused by radiation from space, the decay of natural *radioisotopes, and fallout from the testing of nuclear weapons in the 1960s.

bacterium (pl. bacteria) A type of *micro-organism consisting of only one *cell. There is no *nucleus, so the *chromosomes are free in the *cytoplasm. Bacteria reproduce by *asexual reproduction. They cause the *decay of dead plants and animals and so help to recycle materials. *Toxins from bacteria that are *parasites cause diseases in plants and animals.

balance (in balance, balanced) The state of non-movement in an object, resulting from equal and opposite forces acting on it. A see-saw is balanced when the clockwise and anticlockwise *moments are equal.   ✍

balanced diet For animals, the right amounts and types of food for healthy living and growth. (see *nutrient)

bar chart (bar graph) A diagram showing quantities as (usually vertical) columns. The height of each column is proportional to the size of the quantity.

bar code A pattern of thick and thin lines containing information that can be read by a *computer. Supermarkets use bar codes on food packets to calculate bills and control stock.

barometer A device used to measure *atmospheric pressure. A mercury barometer is a long upright tube filled with mercury that dips into a container of

Balance

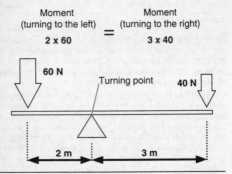

Moment (turning to the left)
2 x 60

=

Moment (turning to the right)
3 x 40

60 N

Turning point

40 N

2 m

3 m

mercury. *Gravitational force pulls down on the mercury in the tube and is balanced by atmospheric pressure pushing up. At sea-level, atmospheric pressure holds up a column of mercury about 760 mm long. An aneroid barometer is a box with flexible sides. If the atmospheric pressure changes, the box changes shape and moves a pointer across a scale.

barytes A mineral containing barium sulphate $BaSO_4$.

basalt A dark glass-like *igneous rock. Basalt is the main type of rock under the *sediment on ocean floors. It is also found on the Moon.

base A substance that can *neutralize an *acid. Like *alkalis, bases can accept protons (H^+ ions) from acids. E.g. zinc carbonate neutralizes hydrochloric acid:

$$ZnCO_3(s) + 2HCl(aq) \rightarrow ZnCl_2(aq) + H_2O(l) + CO_2(g).$$

Bases include metal *oxides, *hydroxides, *carbonates, *hydrogencarbonates, and *ammonia.

base code see *genetic code

base metal A common and fairly cheap metal that tarnishes or corrodes during use. Examples are *iron and *lead.

BASIC (Beginners All-purpose Symbolic Instruction Code) A type of language used by some types of *computer. It is written in familiar English words that can be combined to make up *programs.

battery Two or more electrochemical *cells connected together. A 6-*volt battery is made up from four 1.5-volt cells connected in *series. (NB a single cell is commonly (but wrongly) called a battery.)

bauxite An *ore used in the production of *aluminium. It contains aluminium oxide Al_2O_3.

beam 1. A group of *rays moving together in the same direction. E.g. a small hole in a curtain allows a narrow beam of sunlight into a room. 2. A narrow stream of moving particles (e.g. *alpha radiation).

bearing A device that makes less the *friction between fixed and rotating parts in a *machine. A ball-bearing contains hard steel spheres that roll between the fixed and rotating parts. A bicycle wheel has a ball-bearing at each end of the *axle that it rotates around.

becquerel (symbol Bq) The *SI unit of radioactivity. 1 Bq is the activity of a *radioisotope decaying at the rate of 1 nuclear disintegration per second.

benzene A liquid *hydrocarbon and an *arene. Benzene is made from *crude oil. Its formula is C_6H_6 with the carbon atoms arranged in a hexagonal ring.

Bearing

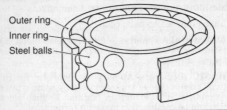

Outer ring
Inner ring
Steel balls

Although *unsaturated, it takes part in *substitution reactions.

beryllium (symbol Be) A grey *metal *element that is a member of group II in the *periodic table (the *alkaline earth metals). Beryllium is extracted by *electrolysis of molten *salts. The metal and its compounds are poisonous. It is used in making *ceramic materials and as a moderator in *nuclear reactors.

beta particle An *electron ejected from the nucleus of a *radioisotope. It is formed when a *neutron decays into a *proton and an electron. (see *nuclear radiation)

beta radiation A stream of *beta particles moving at high speed (up to 9/10 of the speed of light). Its path is strongly bent when travelling close to a magnet or an electrically charged object. Beta radiation can be stopped by a sheet of aluminium 5 mm thick.

biennial A type of plant (usually *herbaceous) that lives for more than one year but less than two. It grows from a seed and stores food in roots or shoots during the first year. In the second year, it produces flowers and seeds, and then dies. Carrots and wallflowers are examples of biennial plants. (see *annual *perennial)

big-bang theory A theory that the history of the Universe started about 16 000 million years ago with the explosion of a fist-sized ball of superdense matter. *Galaxies formed later from the matter flung outwards by this explosion.

bile An alkaline liquid made in the *liver of vertebrates. Bile helps digestive juices from the *pancreas to work. It flows into the *duodenum, where it neutralizes *stomach acid in the partly-digested food. It also breaks fats into droplets. (see *gall bladder)

bimetal strip A bar made from two different metals fixed firmly together side by side. Heating a bimetal strip causes it to bend because the two metals expand by different amounts. Bimetal strips are used in *thermostats and in some types of *thermometer.

binary code The method of writing all numbers using only two *digits, 0 and 1. E.g. the decimal number 5 is 101 in binary code. It is used in *computers. (see *byte)

binary fission A type of *asexual reproduction used by *unicellular organisms. A cell makes a copy of its own *DNA and then splits into two identical daughter cells. (see *mitosis *meiosis)

binocular vision The ability to focus on an object with both eyes at the same time. Animals with forward-facing eyes (owls, humans) have binocular vision. It allows three-dimensional vision and helps the brain to judge distances.

biochemistry The study of the chemical reactions that happen in living organisms.

biodegradable Describing waste materials that can be broken down by *decomposer organisms.

Plant and animal materials such as paper, waste food, and sewage are biodegradable. Many man-made materials such as plastics and glass are non-biodegradable.

biofuel A fuel that is made from a renewable *resource and is an alternative to *fossil fuels. Firewood, *biogas, and *gasohol are biofuels.

biogas A gas that is a *biofuel made from waste materials from plants and animals. It is a mixture of carbon dioxide and methane. Biogas is given off as the waste materials *decay.

biological control The control of pests without using poisonous *herbicides or *insecticides. Examples of biological control methods include breeding disease-resistant plants and releasing natural enemies of the pest.

biological indicator An organism that is very sensitive to *factors in its *environment. *Lichens can be used as biological indicators because they are sensitive to sulphur dioxide in the air. The numbers and types of lichen in an area show the level of air pollution.

biological oxygen demand (BOD) The amount of oxygen used by microbes that decompose organic waste in water. BOD shows the amount of oxygen available for fish and other large organisms and is a measure of some types of water *pollution.

biology The study of all living organisms.

biomass The total mass of a particular type of organism living in a particular place. E.g. the biomass of goldfish living in a pond; the biomass of all the trees in the world.

biome A very large *community of plants and animals covering a large area of the Earth. Examples of biomes are tropical rain forest, desert, and savannah grassland.

biosphere see *ecosystem

biotechnology Using organisms to make or alter materials used in medicines and food and in industry. Biotechnology uses *fungi and *bacteria to help make cheese, wine, beer, yoghurt, *antibiotics, and many other substances. (see *genetic engineering)

biotic factor Each other living organism in the *environment of one particular organism. The biotic factors affecting a rabbit include the plants it eats, its *predators, and all the other rabbits it lives among.

birth-control (contraception) Methods that allow a human couple to decide how many children they have. Birth-control includes the use of *contraceptives, avoiding sexual intercourse when the female is ovulating, and sterilization. Sterilization involves an operation to cut the male *vas deferens or the female *Fallopian tube.

bistable (flip-flop) A type of *logic gate used in *computers. The output of a bistable is switched ON by a current pulse applied to the input. A second input pulse switches the output OFF. Groups of bistables make up memory circuits.

bit The abbreviation for *binary digit*, either 0 or 1 as used in *binary code. In an electronic *computer, 0 corresponds to current flow OFF and 1 corresponds to current flow ON. (see *byte)

bitumen The semi-solid *tar-like residue from the *fractional distillation of *crude oil. Bitumen lakes occur where crude oil has seeped to the surface. (see *tar sand)

black hole In astronomy, an object that contains so much matter that its own *gravitational force has made it collapse in on itself and become very small and superdense. The gravitational force is so great that even light cannot escape from a black hole.

bladder 1. A hollow *organ found in plants or animals. It is used to contain a gas or a liquid. 2. The urinary bladder is used by vertebrates for storing *urine. It is a hollow organ with elastic walls.

blast-furnace An oven like a tall chimney used for *smelting iron *ore to make *pig iron. Iron ore, *limestone, and *coke are added at the top of the furnace. A blast of hot air raises the temperature to 1400°C and the following reactions happen. (1) Coke burns: $C+O_2 \rightarrow CO_2$. (2) Limestone decomposes: $CaCO_3 \rightarrow CaO+CO_2$. (3) Carbon monoxide produced: $C+CO_2 \rightarrow 2CO$. (4) Iron oxide in ore reduced: $Fe_2O_3+3CO \rightarrow 2Fe+3CO_2$. (5) Liquid *slag forms: $SiO_2+CaO \rightarrow CaSiO_3$. Molten iron and slag settle at the bottom of the blast-furnace.

blind spot see *eye

blood A liquid that circulates inside the bodies of animals. It consists of *blood cells in liquid *plasma. Blood carries oxygen and food substances to cells in the body. It removes carbon dioxide and other waste substances from the cells. Blood carries *hormones from *endocrine glands to where they are needed. It moves heat energy around the body. *Leucocytes in blood protect the body against microbes that cause disease.

blood cell Any of the cells (corpuscles) contained in blood. They include red blood cells (*erythrocytes) and white blood cells (*leucocytes). White blood cells are colourless and are also found in the *lymph. They include *phagocytes and *lymphocytes.

blood group A human is born with one of four main types of blood. These are group A, group B, group AB, and group O. The groups are named after

Blood group

Human blood groups

Blood group of donor	Blood group of recipient			
	O	**A**	**B**	**AB** Universal recipient
O Universal donor	✓	✓	✓	✓
A	✗	✓	✗	✓
B	✗	✗	✓	✓
AB	✗	✗	✗	✓

 Compatible transfusions

 Incompatible transfusions

the type of *antigen on the red *blood cells. Group O has no antigens. The *plasma in different blood groups can contain *antibodies to these antigens. These antibodies determine which blood groups can be given to a person during a blood transfusion. ✎

blood pressure The pressure of the blood in the main *arteries in an animal. It changes rhythmically as the heart beats. Blood pressure is highest when the heart muscle contracts (systole) and forces blood into the arteries. It is lowest when the heart muscle relaxes (diastole) as the heart fills with blood. (see *vasoconstriction *vasodilation)

blood vessel A tube that carries blood around the body of an animal. Blood vessels include *veins, *arteries, and *capillaries. Venules are small veins and arterioles are small arteries.

BOD Abbreviation for *biological oxygen demand.

boiling A *physical change where bubbles of vapour grow in the liquid, rise to the surface, and burst. It is the most rapid form of *evaporation. The temperature is high enough for the pressure of the vapour in the bubbles to be greater than the pressure of the atmosphere around them.

boiling-point (b.p.) The temperature at which a pure liquid boils. E.g. water, b.p. = 100 °C; *ethanol, b.p. = 74 °C at *standard atmospheric pressure. Boiling-point is a *physical property used to identify a liquid. Raising the applied atmospheric pressure or adding a non-*volatile *solute (e.g. salt) raises the b.p. of a liquid.

bond see *chemical bond

bone The hard tissue that makes up the *skeleton of a vertebrate. Bone contains calcium *salts and fibres of *collagen. The salts give hardness and the collagen gives strength to the bone. Separate bones that move in *joints (e.g. arm bones and leg bones) have a layer of slippery *cartilage at their ends. Bones are hollow and contain *marrow.

borax The common name for sodium tetraborate $Na_2B_4O_7.10H_2O$, a colourless crystalline solid. It is used to make *antiseptic eyewash, *fungicides, flame retardants, and *catalysts.

boron (symbol B) An *element that is a *metalloid and a member of group III of the *periodic table. It is used in the control rods of *nuclear reactors.

botany The study of plants.

Bowman's capsule see *kidney

Boyle's law A *law that states: 'The volume (V) of a fixed mass of gas at a constant temperature is inversely proportional to its pressure (P).' In symbols, PV = constant. (see *gas laws)

b.p. Abbreviation for *boiling-point.

brain In *vertebrates, an *organ made up from nerve tissue and contained in the skull. It receives information from *sense organs, makes decisions, and controls muscles in the body. The brain connects to the top of the *spinal cord and has three main parts: the *cerebrum, the *cerebellum, and the

Brain

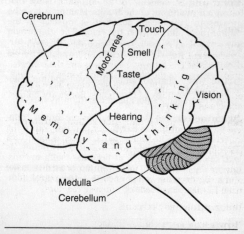

*medulla. The outer part of the cerebrum is called the cortex. ✍

brake A device that slows the speed of a rotating wheel in a vehicle or machine. Hard material (the brake pad, shoe, or block) presses against a revolving steel disc or drum attached to the wheel. *Friction between the material and the revolving part changes *kinetic energy into heat energy, slowing the wheel.

brass A yellow *alloy that contains *copper and *zinc. It is used to make plumbing fittings and machine parts.

brazing solder see *solder

breathing Movements of an animal's body that pump air in and out of the *lungs. Mammals breathe in (inspire, inhale) by contracting muscles in the *diaphragm and between the *ribs. The diaphragm lowers, the ribs rise, and the lungs fill with air. Mammals breathe out (expire, exhale) by relaxing the muscles.

brine Water that contains dissolved sodium chloride (common salt). Electrolysis of brine in industry gives *hydrogen, *chlorine, and sodium hydroxide.

bromine (symbol Br) A *volatile red liquid *nonmetal *element that is a member of group VII in the *periodic table (the *halogens). It exists as molecules Br_2 and is poisonous and reactive. Bromine is extracted from solutions of bromide *salts by *displacement with *chlorine. Bromine is used to make *organic bromine compounds (e.g. the petrol additive 1,2–dibromoethane).

bronchiole see *lung

bronchus see *lung

bronze A yellow *alloy that contains *copper and *tin. It is used to make machine parts and bells.

Brownian motion The random movement of particles of smoke in air when viewed through a microscope. The movement is caused by the smoke particles being constantly bombarded by the invisible molecules that make up air. Brownian motion can also be seen with pollen in water.

brush Part of an *electric motor, *generator, or *alternator. It is used to conduct an *electric current between fixed and rotating parts. A brush is a piece of carbon (*graphite) that is pressed against the rotating *commutator or *slip ring.

bud 1. Part of a *plant that grows into a leaf or a flower. Buds grow on the *stems of plants. 2. Offspring growing on the side of some simple animals. The growth of buds is called budding and is a type of *asexual reproduction.

buoyancy The upward force on a body immersed in a fluid. This force is equal to the *weight of the fluid displaced by the body. A boat floats in water because the upthrust is large enough to support the weight of the boat. An object will float if its *density is less than that of the fluid.

burette A piece of apparatus used to measure out accurately small volumes of liquid. It consists of a long narrow transparent tube fitted with a tap. The tube has a scale (usually 0–50 cm³) that can be accurately read to 0.05 cm³.

burn see *combustion

butane A volatile liquid (boiling-point 0 °C) that is a *hydrocarbon and an *alkane. Its formula is C_4H_{10} and there are two *isomers. Butane is made by *fractional distillation of *crude oil. It is used as a fuel

Burette

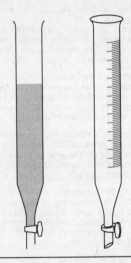

('bottled gas') and in the manufacture of synthetic rubber.

butanol see *alcohol

by-product A substance formed during a *chemical reaction at the same time as the main *product. Some by-products are useful, e.g. platinum is recovered during the purification of copper by *electrolysis. Many industrial by-products can be the cause of *pollution.

byte A group of *bits, usually 8, used to express numbers, letters, or instructions in a *digital *computer. The size of a computer's memory and the capacity of a *disk are measured in bytes (kilobytes or megabytes).

C

caecum In vertebrates, a part of the *alimentary canal at the end of the *ileum. The caecum is a pouch at the join between the small and the large *intestines. Herbivores have a large caecum containing *bacteria that help to digest *cellulose. Carnivores have a very small caecum. (see *appendix)

calcite A *mineral consisting of calcium carbonate. It is present in many types of *limestone and *marble.

calcium (symbol Ca) A soft grey *metal *element that is a member of group II in the *periodic table (the *alkaline earth metals). It is extracted by *electrolysis of molten calcium chloride. Many *minerals contain calcium (e.g. *limestone, *marble, *gypsum). It is an *essential element for living organisms. Calcium is used in the extraction of *uranium.

calcium carbonate (formula $CaCO_3$) A *salt that is found in nature as *limestone, *chalk, and *marble. It is insoluble in water and gives carbon dioxide when mixed with nitric acid.

calibrate To add a *scale to a measuring instrument or to adjust a measuring instrument so that it is accurate. A thermometer is calibrated by putting it in melting ice to fix the 0 °C point and then in steam (at *standard atmospheric pressure) to fix the 100 °C point.

calorimeter A piece of equipment used to measure changes in *heat energy. The change that liberates or absorbs heat energy takes place inside the calorimeter, which is insulated from its surround-

Cam

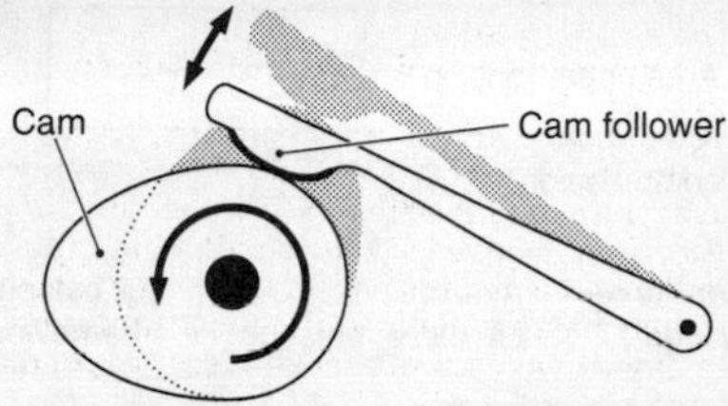

Cam

Cam follower

---

ings. The change in heat energy is calculated by multiplying the temperature increase or decrease by the *heat capacity of the calorimeter.

calyx The lower part of a *flower that originally formed the outside of the flower bud. The calyx is made up from separate sepals which are usually green and leaf-like.

cam An unevenly shaped wheel that rotates and makes an object pressing against its side move back and forth. Cams open and close the *valves in a petrol *engine.

cancer A type of disease caused by *mutations in the cells of an organism. Cells change and multiply, forming a ball of cells called a growth or tumour. A malignant tumour grows rapidly and spreads throughout the body, damaging or destroying vital organs. A benign tumour grows slowly in one place, causing little damage.

canine A type of *tooth used for piercing and tearing food. There are two canines in each jaw, between the *incisors and *premolars. *Herbivores do not have canine teeth.

Cantilever

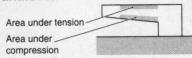

Area under tension
Area under compression

Cantilever bridge

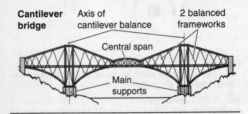

Axis of cantilever balance — 2 balanced frameworks

Central span

Main supports

---

cantilever A beam supported at one end only. The free end of a cantilever is held up by *tension or *compression forces. A bracket under a shelf is a simple cantilever. ✍

capacitor A device that can store an electric *charge. Most capacitors consist of two sheets of metal foil separated by an insulator and rolled into a tube. Charge builds up on the sheets when a *voltage is connected across them. ✍

capillary In mammals and other animals, a microscopic *blood vessel with very thin walls. Networks of capillaries pass around all the cells in the body. *Tissue fluid and *leucocytes escape from capillaries and help to nourish and protect the cells. Capillaries receive blood from *arteries and pass it on to *veins.

Capacitor

Symbols:

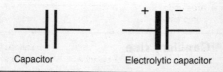

Capacitor   Electrolytic capacitor

capillary rise (capillarity) The ability of water and other liquids to rise up inside a tube with a small diameter. It is caused by forces of attraction between the liquid and the walls of the tube. The narrower the tube, the greater the capillary rise.

carbohydrate A type of solid substance made up from carbon, hydrogen, and oxygen. Carbohydrates are the main source of energy for living things. They are made by plants as a result of *photosynthesis. *Sugars, *starch, *cellulose, and *glycogen are carbohydrates.

carbon (symbol C) A non-metallic *element that is a member of group IV of the *periodic table. Pure carbon can exist as two *allotropes, *graphite and *diamond. (see *organic compound)

carbonate An *ion and an acid *radical with the formula CO_3^{2-}. Carbonates form *salts, e.g. sodium carbonate Na_2CO_3. All carbonate salts give carbon dioxide when added to nitric acid. Most carbonate salts (except *alkali metal carbonates) are insoluble in water and give off carbon dioxide when heated.

carbon cycle The movement of carbon between *carbon dioxide in the atmosphere, living organisms, and *fossil fuels. Plants grow and live by taking in carbon dioxide during *photosynthesis. Most organisms give out carbon dioxide during *respiration. *Combustion returns carbon dioxide to the atmosphere.

Capillary rise

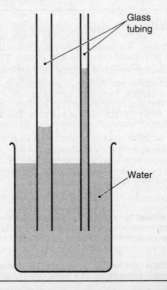

Glass tubing

Water

carbon dioxide A gas with the formula CO_2 that makes up 0.03% of the Earth's atmosphere and is the result of *combustion. It is given out by living *organisms during *respiration and is used by plants during *photosynthesis. It is used in fire extinguishers and *dry ice. (see *yeast *carbon cycle)

carbon-14 dating Estimating the age of carbon-containing remains by measuring the amount of the *radioisotope carbon-14 (^{14}C). The concentration of ^{14}C in the environment remains constant. Its rate of decay is matched by its rate of formation from nitrogen in the upper atmosphere. The concentration of ^{14}C in an organism remains constant while it is alive and taking in (carbon-based) food. Radioactive *decay causes the concentration of ^{14}C to fall steadily after death.

carbon monoxide A gas with the formula CO that results from incomplete *combustion. It is present in car exhaust fumes and cigarette smoke. Carbon monoxide is poisonous. It stops the uptake of oxygen by *haemoglobin in the blood by forming carboxyhaemoglobin.

carboxylic acid A type of *organic compound that contains the carboxyl group $-COOH$. The simplest carboxylic acids (*fatty acids) form a series of liquids: methanoic acid $HCOOH$; ethanoic acid CH_3COOH; propanoic acid C_2H_5COOH; butanoic acid C_3H_7COOH. These dissolve in water to form weak acids.

carcinogen Anything that can cause *cancer. Tobacco smoke, some chemicals, *X-rays, and *nuclear radiation can act as carcinogens.

carnassial A type of *tooth found in *carnivores and used for slicing flesh. Carnassial teeth are *premolars or *molars that have sharp cutting edges.

carnivore Any animal that eats meat, e.g. cat, shark, dragon fly *larvae. The *carnivores* are a group (*order) of *mammals that eat meat, e.g. cat, wolf, walrus. Members of this order have powerful jaws and large *canine and *carnassial teeth.

carpal In humans and many other land *vertebrates, one of the bones that make up the wrist. (see *skeleton)

carpel see *flower

carrier A person that has an *allele for a defect that is masked by a normal dominant allele. The person does not suffer from the disease, but may pass it on to their children. (see *haemophilia *sex linkage)

carrier wave A continuous radio signal with a fixed *frequency and *amplitude that is broadcast by a radio transmitter. For the signal to carry information, the carrier wave is switched on and off (Morse code and data signals) or it is changed by *modulation (sound, vision, and data signals).

cartilage (gristle) A tough but flexible type of *tissue found in *vertebrates. The skeletons of some fish (e.g. sharks) are made from cartilage. Animals with bony skeletons have cartilage at the ends of *bones that make up *joints. The end of the nose and the pinna of the *ear also contain cartilage.

cast iron see *pig iron

catalase An enzyme in the liver that decomposes poisonous *hydrogen peroxide.

catalyst A substance that alters the rate of a chemical reaction without itself being used up. Each catalyst affects only certain reactions. E.g. copper ions speed up the production of hydrogen from zinc and sulphuric acid. (see *Haber process *contact process).

cathode An *electrode that has a negative *charge.

cathode-ray tube (CRT) An electronic device that uses a beam of *electrons to produce a spot of light on a fluorescent screen. There is a CRT in most TV sets, using electromagnets to deflect the beam so that it scans the screen. The picture is made up from a series of horizontal lines. The beam in an *oscilloscope CRT is deflected by electric *fields between two sets of plates.

cation An *ion with a positive *charge, attracted to the *cathode during *electrolysis. Cations include metal ions (e.g. sodium Na^+ and aluminium Al^{3+}) and molecular ions (e.g. ammonium NH_4^+ and *hydroxonium H_3O^+).

cell 1. A living cell is the smallest part of which all *organisms are made. The average diameter of a cell is 0.01–0.1 mm. Cells are made up from *protoplasm enclosed by the *cell membrane. Protoplasm consists of the *nucleus and its surrounding *cytoplasm. The cytoplasm contains *organelles such as *mitochondria and *ribosomes. Small *vacuoles may be present in animal cells. Plant cells have *chloroplasts, large vacuoles, and a *cell wall surrounding the cell membrane. 2. An electrochemical cell usually consists of two solid *electrodes in contact with a liquid *electrolyte. During *electrolysis or *electroplating, an electric current passes through the electrolyte between the electrodes. This causes chemical changes in the cell. In a voltaic cell, *chemical reactions between the electrodes and the electrolyte produce an electric current. (see *primary cell *secondary cell)   ✍

cell membrane The outer covering of a living *cell. It is a *semi-permeable membrane made up from *fat and *protein. The cell membrane controls the flow of substances in and out of the cell.

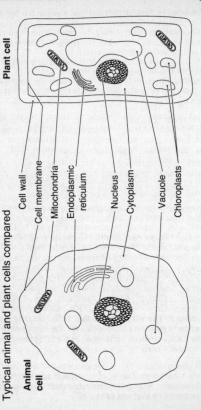

Cell

Typical animal and plant cells compared

Plant cell

Animal cell

- Cell wall
- Cell membrane
- Mitochondria
- Endoplasmic reticulum
- Nucleus
- Cytoplasm
- Vacuole
- Chloroplasts

44

cellulose A *carbohydrate that makes up the *cell walls of plants. Cellulose is a *polymer made up from *glucose molecules. Humans cannot digest cellulose. It is the main part of *dietary fibre. Cotton and artificial silk are made of cellulose.

cell wall The outer covering of *plant *cells and *bacteria cells. It is fairly rigid and surrounds the *cell membrane. This maintains the shape of the cell. The cell walls of plant cells contain *cellulose.

Celsius scale A scale for measuring *temperature. 1 degree Celsius (1 °C) is equal in magnitude to 1 K (1 *kelvin).

cement 1. Bony material that holds the root of a *tooth in the jawbone. 2. Portland cement is a grey powder made by roasting and grinding a mixture of *limestone and *clay. It sets into a hard mass when mixed with water. (see *mortar *concrete)

centigrade scale The old name for the *Celsius scale, used for *temperature measurement.

central nervous system (CNS) In vertebrate animals, the part of the *nervous system made up from the *brain and the *spinal cord.

central processing unit (CPU) The part of a *computer that contains the *program and which processes *data flowing through it. The CPU is made up from many thousands of *logic gates.

centre of gravity The point at which the whole weight of an object seems to act. The total result of *gravitational force acting on each of the particles in the object seems to be concentrated at the centre of gravity.

centre of mass The point at which the whole mass of an object seems to be concentrated. It is the same point as the *centre of gravity.

Centre of gravity

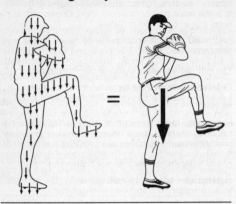

---

centrifugal force see *centripetal force

centrifuge A machine used to separate the solid from the liquid in a *suspension. Test-tubes filled with the suspension are spun round at high speed. The rapid rotation flings the solid to the bottom of the tubes.

centripetal force The *force which makes an object move on a circular path. As the object moves, centripetal force pulls it towards the centre of the circle. *Gravitational force is the centripetal force that keeps a planet orbiting around the Sun. An out-of-date method of explaining circular motion used the idea of a *centrifugal force* acting *outwards* from the centre of the circle. A planet was said to stay in its orbit because the outward centrifugal force was balanced by the inward gravitational force. We now simply say that the gravitational force is the centripetal force. ✍

ceramic A hard stony or glassy *inorganic material, usually with a very high melting-point, e.g. pottery, enamels.

cerebellum A part of the *brain in *vertebrates. The human cerebellum helps to co-ordinate muscles and make them work together.

cerebrum The largest and most highly developed part of the *brain in *vertebrates. The human cerebrum contains 10 000 million nerve cells. It receives information from sense organs and controls voluntary movement and memory and thinking.

cervix In female mammals, the narrow neck of the *uterus where it connects to the top end of the *vagina. (see *sexual reproduction)

Centripetal force

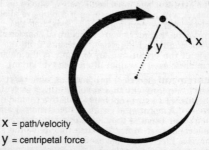

X = path/velocity

y = centripetal force

CFC Abbreviation for *chlorofluorocarbon.

chain reaction A *reaction producing products that cause the reaction to continue. *Nuclear fission of a uranium-235 atom releases three *neutrons which cause the fission of three more atoms, and so on. An uncontrolled chain reaction leads to a nuclear explosion. (see *nuclear reactor)

chalk A white *sedimentary rock made of the skeletons or shells of microscopic sea creatures. It is mostly calcium carbonate. NB sticks of chalk used for drawing are made from *gypsum, calcium sulphate.

characteristic Something we observe about an organism that helps us to recognize and name it. Inherited characteristics include feathers on birds and walking upright on two legs in human beings. *Genes on *chromosomes are responsible for passing inherited characteristics from parents to offspring (see *phenotype). Each characteristic of an organism is controlled by specific genes called *alleles. Blue eye colour is a recessive characteristic controlled by a pair of *recessive alleles. A baby will only have blue eyes if it has inherited a pair of alleles for blue eyes. It will have brown eyes if a blue-eye allele is paired with a *dominant brown-eye allele (see *heterozygous *homozygous). An acquired characteristic is not controlled by genes. It results from the lifestyle of the organism. An example is the enlarged arm muscles of a weight-lifter. (see *variation)

charge An electrical property of some particles, used to explain why they repel or attract each other. There are two sorts of charge, positive (+) and negative (−). A *proton (+) and an *electron (−) attract each other because they have opposite charges. Protons repel each other because they have the same charges. Electrons repel each other for the same reason.

Charles' law A *law that states: 'The volume (V) of a fixed mass of gas at a constant pressure is proportional to the kelvin temperature (T). In symbols, V = constant×T. (see *gas laws)

chemical A single *pure substance that is either an *element or a *compound.

chemical bond The *electrostatic force of attraction that holds *particles together in a *compound. Chemical bonds include *ionic (electrovalent), *covalent, and *metallic bonds.

chemical change see *chemical reaction

chemical energy *Energy that is within the *chemical bonds of a substance. Chemical energy is changed into heat and light energy during the *combustion of fuels. Chemical energy is also contained in food.

chemical equation The *formulas of *reactants and *products written down to show how they take part in a *chemical reaction. The reactants are placed to the left of the products; an arrow that stands for 'reacts to give' connects reactants to products. E.g. the reaction of magnesium carbonate with hydrochloric acid:

$$MgCO_3(s) + 2HCl(aq) \rightarrow CO_2(g) + H_2O(l) + MgCl_2(aq).$$

1 *mole of solid magnesium carbonate reacts with 2 moles of hydrochloric acid solution to give 1 mole of carbon dioxide gas, 1 mole of liquid water, and 1 mole of magnesium chloride solution.

chemical equilibrium In a reversible *physical change or *reversible reaction, the state where the amounts of all the substances present are steady. E.g. in a sealed container, water is in equilibrium with its vapour; in the *Haber process, hydrogen and nitrogen are in equilibrium with ammonia. Equilib-

Equilibrium

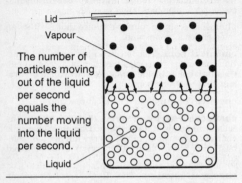

Lid

Vapour

The number of
particles moving
out of the liquid
per second
equals the
number moving
into the liquid
per second.

Liquid

rium results when the rate of the forward reaction
equals the rate of the backward reaction.

chemical property Any one of the chemical
changes (*chemical reactions) that can happen to a
substance. E.g. a chemical property of *sulphuric
acid is that it can *neutralize *bases.

chemical reaction A change which results in new
substances (products) being formed from starting
substances (reactants). The masses of the reactants
and products in a reaction are in a fixed ratio to each
other. E.g. 4 g of hydrogen reacts with 32 g of oxygen
to make 36 g of water—the ratio of hydrogen to oxygen to water by mass is 1:8:9. A chemical change is
usually difficult to reverse. Energy is given out or
taken in during a chemical reaction.

chemical symbol An abbreviation for the name of an *element. Chemical symbols are used for writing the *formulae for *compounds and *chemical equations. Some elements have symbols based on their Latin names, e.g. iron Fe—*ferrum*; potassium K—*kalium*.

chemistry The study of the *elements and the compounds they form.

chip A small crystal of specially treated silicon, usually about 2 mm square, used in the manufacture of *integrated circuits.

chitin A substance that strengthens the exoskeletons of *arthropods and the jaws of *annelids. It also occurs in *fungi.

chlorination 1. A *chemical reaction where a compound joins with chlorine. 2. *Disinfecting water by adding chlorine and so making it safe to drink or bathe in.

chlorine (symbol Cl) A pale greenish-yellow gaseous *non-metal *element that is a member of group VII in the *periodic table (the *halogens). It exists as molecules Cl_2, and is poisonous and highly reactive. Chlorine is extracted by *electrolysis of sodium chloride, either molten or as a solution. It is used to make *hydrochloric acid, for bleaching, for the *chlorination of water, and to make *organic chlorine compounds, e.g. *PVC, *insecticides, *herbicides, *solvents.

chlorofluorocarbon (CFC) A type of volatile liquid compound made up of carbon, fluorine, and chlorine atoms. Chlorofluorocarbons are used in refrigerators and in the manufacture of some aerosols and plastics. (see *ozone layer)

chlorophyll The substance that gives a *plant its green colour. It is contained in the *chloroplasts inside plant cells. Chlorophyll absorbs light energy which the plant uses to help make its own food. (see *photosynthesis)

chloroplast An *organelle containing *chlorophyll inside a plant *cell. *Photosynthesis takes place inside chloroplasts.

cholesterol A substance found mostly in animal tissues and blood. It is absorbed from food or made by the liver. Cholesterol is used to make cell membranes, part of *bile, *vitamin D, and many *hormones. Too much cholesterol in the blood causes solid deposits that can block arteries.

choroid see *eye

chromatid A pair of joined *chromosomes that results when a single chromosome copies itself during *meiosis or *mitosis.

chromatography A process used to separate mixtures of dissolved substances. E.g. in paper chromatography, a spot of the mixture is placed on absorbent paper. Pure *solvent flows across the spot. The different dissolved substances are carried along at different speeds, depending on how well they dissolve in the solvent or are absorbed by the paper. The substances separate into bands.

chromium (symbol Cr) A hard silvery *transition metal *element. Chromium is used in making stainless *steel. It can be *electroplated on to other metals to give a hard and shiny protective coat.

chromosome A threadlike structure found in the nucleus of a plant or animal *cell. Chromosomes are made up from *genes which control the appearance of inherited characteristics. They are arranged in pairs during *mitosis and *meiosis. Each *species has its own particular number and type of chromosomes. Human cells have 46 chromosomes arranged

Circulation

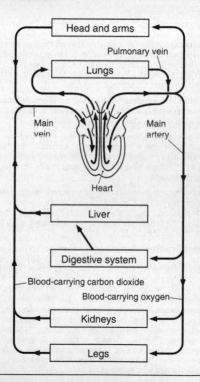

chrysalis   54

in 23 pairs. The *gametes in an organism contain
half the usual number of chromosomes.

chrysalis see *pupa

ciliary muscle see *eye

cilium (pl. cilia) A hair-like thread that grows from
the surface of some types of *cell. Groups of cilia beat
rhythmically. Cilia propel some microbes (e.g. *Para-
mecium*) through a liquid. Larger animals use cilia to
move internal fluids (e.g. *mucus).

circuit A complete path made up from conductors
through which an *electric current can flow. (see
*parallel connection *series connection)

circulation A *circulatory system* that carries substances around a living body. In animals, the *heart,
*blood, and *blood vessels make up the blood circu-
lation. *Lymph and lymph vessels make up the lym-
phatic system.   ✍

class A group of living organisms with similar
*characteristics. Several classes make up a *phylum
or *division. Lions are in a class called the *mammals, which also includes dogs, walruses, and mice.
(see *classification)

classification A method of sorting living organisms into different groups. Each group contains
organisms with similar *characteristics. A simple
form of classification has two main groups, called
the *plant *kingdom and the *animal kingdom. The
modern form of classification has five main kingdoms: plants, animals, *fungi, *monera, and *protoctista. Each kingdom is broken into smaller and
smaller groups in the order kingdom, *phylum or
*division, *class, *order, *family, *genus, and
*species. Members of a larger group (e.g. phylum)
have fewer similarities than the members of a
smaller group (e.g. genus).

Species	lion
Genus	lion tiger leopard
Family (the cat family)	lion tiger leopard jaguar domestic cat
Order (the *carnivores)	lion tiger leopard jaguar domestic cat dog bear walrus
Class (the mammals)	lion tiger leopard jaguar domestic cat dog bear walrus bat rabbit hippopotamus
Phylum (the vertebrates)	lion tiger leopard jaguar domestic cat dog bear walrus bat rabbit hippopotamus bird frog snake
Kingdom (the animals)	lion tiger leopard jaguar domestic cat dog bear walrus bat rabbit hippopotamus bird frog snake worm jellyfish shark spider wasp snail

clavicle In humans, the collar-bone. It joins the shoulder-blade (*scapula) to the breastbone (*sternum). (see *skeleton)

clay A material made from particles of *rock (formed by *weathering) that are usually smaller than 0.01 mm in diameter. The *minerals in clay (mainly aluminium silicate) help it to hold water and make it soft, *plastic, and *impermeable.

climate The range in temperature, humidity, and rainfall for a part of the Earth's surface. There are four main types of climate: polar (e.g. Antarctica), temperate (e.g. Europe), subtropical (e.g. North Africa), and tropical (e.g. West Indies).

climatic factor Part of the *environment that affects the life-style of an organism living in a *habitat. Climatic factors include *temperature, rainfall, wind, *humidity, and light intensity.

clinostat A machine used to study the effect of a *tropism (geotropism or phototropism) in plants. It rotates a whole seedling while it grows and so cancels out the effect of gravity or light.

clitoris A sensitive rod of tissue in female mammals lying above the opening to the *vagina.

cloaca The opening at the end of the *alimentary canal in all vertebrates, excluding mammals. The cloaca also connects with tubes from the reproductive organs and the *kidneys.

clone Two or more organisms that have identical *genes. The organisms come from a single parent by *asexual reproduction, e.g. *cuttings taken from one plant.

clot A solid protective coat that forms at a wound. *Platelets in the blood trigger the release of *enzymes from injured tissues. The enzymes control the formation of tangled fibres that trap blood cells and form a clot.

CNS Abbreviation for *central nervous system.

coagulate Large particles in a solution joining together to form a solid. Boiling an egg coagulates protein in the yolk and egg-white. River deltas form when silt particles coagulate as they meet sea water.

coal A solid black *fossil fuel, formed from plants that grew over 200 million years ago. It contains over 90% of carbon. Most coal is burned as a *fuel in *power-stations. (see *coke)

cocaine A stimulant *drug obtained from the leaves of the coca bush. It is also used to make local *anaesthetics. 'Crack' is a highly addictive form of cocaine that can be smoked.

cochlea see *ear

co-dominance The condition in an organism when a *characteristic results from the *expression of two *alleles which are both *dominant. Neither allele is

dominant to the other. The human AB blood group is
the result of two alleles A and B being expressed.

coelenterate A type of *invertebrate animal. Examples are *Hydra*, jellyfish, and sea anemones. Coelenterates have soft bag-shaped bodies.

cog Each of the teeth that stick out from the sides of
a *gear wheel.

coherent light *Light made up from *waves that
are all of the same *frequency and that are all vibrating in step with each other. A *laser gives out a beam
of coherent light.

coke An impure form of carbon made by heating
*coal in the absence of air. It is mainly used in *blast-
furnaces to make *pig iron.

cold-blooded see *poikilotherm

collagen A tough substance found in the skin, tendons, and bones of animals. Collagen is a protein that
forms strong fibres which do not stretch easily.

colloid A substance (e.g. jelly, wallpaper paste)
made up from *particles dispersed in a liquid. The
particles are larger than the *molecules or *ions in a
*solution, but smaller than the particles in a *sus-
pension. *Filtration does not separate the particles
from the liquid. If the particles are gas bubbles, the
colloid is a foam.

colon The main part of the large *intestine between
the *caecum and the *rectum. It absorbs water and
mineral salts from food that has passed through the
small intestine.

colour-blindness A person with colour-blindness
cannot tell the difference between certain colours.
More males than females suffer from it. Red–green
colour-blindness is the most common and is caused
by defects in the retina of the *eye.

combustion (burning) An *exothermic *chemical reaction between a substance and oxygen, giving out heat and light. E.g. the combustion of *methane in air:

$$CH_4(g)+_2O_2(g) \rightarrow CO_2(g)+2H_2O(g).$$

combustion chamber The place inside an *internal combustion engine or *rocket engine where the *fuel burns.

comet A small body (diameter about 1 km) made of ice, dust, and gas that moves around the Sun. The *orbit of a comet is extremely elliptical, with a *period usually between 70 and 100000 years. A comet has its maximum speed when closest to the Sun. The *solar wind causes the comet to develop a 'tail' up to 10000000 km long that always points away from the Sun.

commensalism The relationship between two organisms from different *species that live together and share the same food. One of the species benefits from this relationship, the other is not affected. Sucker fish (remora) cling to the bellies of sharks and receive protection and scraps of food.

community A group of organisms living together in a *habitat. They have an effect on each other and are linked by a *food web. The number of plants or animals in each *species stays fairly constant. (see *succession *biome *ecosystem)

commutator In some *electric motors and *generators, a part that is connected to each end of the coil of wire that spins with the *rotor. Fixed carbon *brushes press against the commutator and conduct electricity through the moving coil.

composite A material made from two or more separate materials, combining their properties. E.g. *glass-reinforced plastic (GRP) is tough and flexible. Wood is a natural composite.

compost Plant remains collected together and allowed to *rot. Compost is added to the soil to increase the *humus content.

compound A substance that contains two or more different *elements held together by *chemical bonds. (see *ionic compound *covalent compound *organic compound *inorganic compound)

compression A pushing force that squashes something. Sitting on a chair causes a force of compression in each of its legs. Compression of a gas lowers its volume and raises its pressure.

compression ratio In an *internal combustion engine, the comparison between the maximum and the minimum volume of the combustion chamber as the piston falls and rises. The compression ratio for a petrol engine is about 8:1. For a *diesel engine it is about 23:1.

computer An electronic device that takes in and stores information (*data) and processes it to solve problems or to generate a display. A *digital computer usually takes in data from a keyboard, *disk, or tape. The data is manipuated in the *central processing unit according to a stored *program which controls the functioning of the whole computer. The results can be shown on a *visual display unit or be used to control a machine or to drive a *printer. (see *hardware *software *binary code *random access memory *read only memory)

concave lens (diverging lens) A *lens with one or both surfaces curving inwards. Any light *ray arriving parallel to the *axis of a concave lens is deflected (bent) so that its path can be traced back to a point called the principal focus. Rays through the centre of the lens are undeflected. A concave lens always forms an *image that is smaller than the object and is the same way up.   ✍

Concave lens

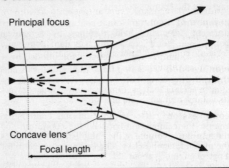

Principal focus

Concave lens

Focal length

---

concentration The amount of a substance contained in a certain volume. E.g. dissolving 10 grams of salt in 50 cm³ of water gives a solution with a concentration of 200 grams per dm³ (200 grams/litre). The concentration of a solution can also be calculated in moles per dm³. Small concentrations of impurities are given as parts per million (ppm). (see *molarity)

concrete A hard rock-like material made when a mixture of *cement, water, sand, and small stones sets. It is used to make buildings and bridges and some road surfaces. Reinforced concrete contains steel rods that resist *tension and stop the concrete from cracking when bearing a heavy load.

condensation reaction A *chemical reaction where two molecules join to form a larger molecule, at the same time giving off a simple molecule such as water H_2O or ammonia NH_3. E.g. a condensation reaction in plants forms *starch when molecules of *glucose join and give off water.

condense To change a *vapour (gas) into a liquid by cooling or by *compression.

condenser A device used to cool a vapour to change it into a liquid. A Liebig condenser consists of a pipe through which the vapour passes, surrounded by a larger pipe through which cold water flows. 🖉

conduction 1. (thermal conduction) *Heat flowing from a high-temperature (hot) part of a substance to a lower-temperature (cooler) part, without the substance itself moving. If part of a *solid is heated, atoms take in energy by vibrating more rapidly. Collisions cause neighbouring atoms to speed up and heat spreads through the solid. Metals have high *conductivity because some electrons can move freely between all the atoms. Heating a metal causes these electrons to speed up and rapidly carry heat to all parts. 2. (electrical conduction) Electric *charge flowing through a *conductor, resulting from a *potential difference applied across the conductor.

Condenser

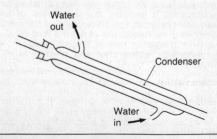

Water out

Condenser

Water in ➡

conductivity A measure of how easily heat (*thermal energy) or an *electric current can pass through a substance. Metals have good conductivity. (see *conduction)

conductor A substance through which *thermal energy (heat) or an *electric current can easily flow. All *metals and the non-metal *graphite are solids that are good conductors because they contain electrons that are free to move from one atom to another. A good electrical conductor has a low *resistance. (see *electrolyte *insulator)

cone cell A type of light-sensitive cell in the retina of the *eye. Cone cells work in bright light and are responsible for colour vision.

conglomerate *Sedimentary rock containing small pebbles.

conjunctiva see *eye

connecting rod In some *machines (e.g. *internal combustion engines), a rod that connects a *piston to a *crankshaft.

conservation The sensible use of the world's *resources with minimum spoiling of the natural environment. It includes: the slowest use of non-renewable resources; the development of *renewable energy sources; careful use of natural food supplies (e.g. fish); *recycling; making the minimum amount of *pollution; looking after and creating wildlife *habitats.

conservation of energy A *law stating that *energy is neither created nor destroyed as it changes from one form to another. E.g. the total amount of heat and light energy given out by a light bulb equals the amount of electrical energy going into the light bulb.

conservation of mass A *law which says that throughout a chemical reaction the quantity of matter stays constant, i.e. the total mass of the *products at the end of the reaction is the same as the total mass of the *reactants at the start. This law does not apply to *nuclear reactions which convert matter into energy.

consumer An organism that obtains its food from plants or animals. Primary consumers are herbivores; they eat plants. Secondary consumers are carnivores that eat herbivores. (see *food chain)

consumer unit A box containing a main switch and fuses, joined to all the electric mains circuits in a house.

contact process The industrial method used to make *sulphuric acid. Sulphur is burnt in air or *sulphide ores are roasted to make *sulphur dioxide. A vanadium oxide catalyst combines sulphur dioxide and oxygen to form sulphur trioxide. This gas is dissolved in sulphuric acid to make oleum (disulphuric acid). This liquid is then added to water to make concentrated sulphuric acid.

continental drift The theory that the Earth's continents formed from the breakup of a single continent (Pangaea) that existed 200 million years ago. Continental drift continues today. E.g. the Atlantic Ocean becomes 3 cm wider each year. (see *plate tectonics)   ✍

contour ploughing Ploughing furrows across sloping ground instead of up and down it. Contour ploughing helps to limit *soil erosion. The direction of the furrows stops rain-water rushing down the hill.

contraceptive Something that prevents pregnancy as a result of sexual intercourse. Common contraceptives include condoms, diaphragms, and pills. A

Continental drift

200 million years ago

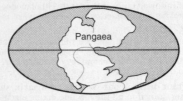

135 million years ago

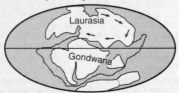

65 million years ago

condom is a thin rubber cover that fits over the penis
and catches sperms. A diaphragm (cap) is a springy
rubber circle that covers the woman's *cervix and
holds a cream that kills sperms. The contraceptive
pill is swallowed daily and contains *hormones that
prevent *ovulation.

contraction The shortening of an object, especially
a muscle. The word *contractions* usually refers to
repeated contractions of muscles in the *uterus wall
as a baby is born.

control experiment An experiment carried out to
ensure that unexpected or unknown variables do not
affect another (main) experiment. E.g. when testing
the effect of a fertilizer on plant growth, a control
experiment will grow the same type of plants under
exactly the same conditions but without fertilizer.
Comparison with the control will eliminate effects
due to the weather.

control rod see *nuclear reactor

convection *Heat flowing from one part of a *fluid
to another by the movement of the fluid itself. E.g.
convection causes hot air to rise from a burning can-
dle because the hot air is less dense than the surrounding colder air. The stream of moving air is
called a convection current.

convex lens (converging lens) A *lens with one or
both surfaces bulging outwards. Any light *ray
arriving parallel to the *axis of a convex lens is
deflected (bent) so that it passes through a point
called the principal focus. Rays through the centre of
the lens are undeflected. Convex lenses in cameras
and eyes form *images that are upside-down. A convex lens works as a magnifying glass when the object
is closer than the *focal length of the lens.   ✍

copper (symbol Cu) A red-brown *transition metal
*element. Copper is extracted by heating its *ore in

Convex lens

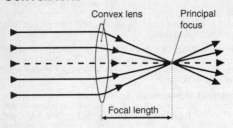

Convex lens
Principal focus
Focal length

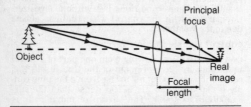

Principal focus
Object
Focal length
Real image

---

air. The impure copper that results is *refined (purified) by *electrolysis. It is very *malleable and *ductile and an excellent conductor of heat and
electricity. It is used to make pipes, electric wire,
and alloys such as *brass and *bronze.

cornea see *eye

corpuscle see *blood cell

corrosion Attack on the surface of a metal by gases
or liquids in contact with it. Corrosion weakens
structures and is commonly the result of water and

oxygen in the air. (see *rusting *sacrificial protection)

cortex In plants, the outer part of a stem or root. In animals, the outer part of an *organ. (see *kidney *brain)

cotyledon A leaf which is part of an *embryo plant inside a *seed. Seeds have one or two cotyledons. The cotyledons in some seeds (e.g. pea, bean) contain a store of food. Cotyledons sometimes grow up out of the ground and form the first leaves on the seedling plant. (see *germination)

coulomb (symbol C) The *SI unit of electrical *charge. It is equal to a current of 1 ampere flowing through a *conductor for 1 second. It is also the total charge carried by approximately 6.24×10^{18} electrons.

covalent bond The *electrostatic force of attraction between two neighbouring nuclei and a shared pair of electrons between them. By sharing electrons, each atom achieves the stable *electronic configuration of a noble gas. E.g. the two atoms in a molecule of chlorine Cl_2 are held together by a cova-

---

Covalent bond

An oxygen atom and two hydrogen atoms

A molecule of water (H_2O)

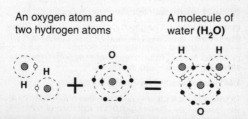

---

lent bond; each atom has the same electronic configuration as argon. The shared pair of electrons may be represented in a formula as Cl:Cl or as Cl–Cl. ✍

covalent compound A *compound formed from non-metal elements only and that consists of molecules. E.g. water H_2O, *polythene. Covalent compounds have low melting- and boiling-points and do not conduct electricity when molten.

Cowper's gland A *gland connected to the *urethra in male mammals. It produces a liquid that mixes with *semen as it passes down the urethra. (see *sexual reproduction)

CPU Abbreviation for *central processing unit.

cracking A process that uses heat to split (*decompose) large *hydrocarbon molecules. Less useful *fractions from *crude oil can be cracked to make *petrol and *diesel oil, together with *ethene and *propene. ✍

cranium The part of the *skull of a vertebrate that contains the brain. (see *skeleton)

crankshaft A part of some machines that converts up-and-down motion into rotary (round and round) motion. The pedals on a bicycle are joined to a simple crankshaft. (see *internal combustion engine)

critical angle see *total internal reflection

crop rotation The yearly changing of crop types growing on a piece of land. Different crops take different nutrients from the soil. Some crops such as peas and beans put nutrients into the soil. Crop rotation helps to stop *soil depletion.

CRT Abbreviation for *cathode-ray tube.

crude oil (petroleum) A dark-coloured *volatile liquid *fossil fuel, made up from a mixture of over 300 different *saturated *hydrocarbons. It was formed

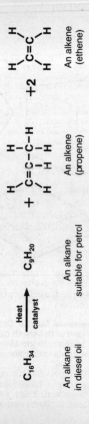

Cracking

$$C_{16}H_{34} \xrightarrow{\text{Heat catalyst}} C_9H_{20} + \begin{array}{c} H \\ | \\ C=C-C-H \\ | \quad | \quad | \\ H \quad H \quad H \end{array} + 2 \begin{array}{c} H \qquad H \\ \diagdown \quad \diagup \\ C=C \\ \diagup \quad \diagdown \\ H \qquad H \end{array}$$

An alkane in diesel oil An alkane suitable for petrol An alkene (propene) An alkene (ethene)

Crystal

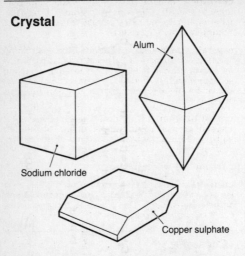

Alum

Sodium chloride

Copper sulphate

---

from plant material that grew over 250 million years ago. (see *oil refinery)

crust Another name for the *lithosphere, the outer (mostly solid) part of the Earth that covers the *mantle. The crust is about 70 km thick under mountains and about 8 km thick under the oceans.

crustacean A type of *arthropod animal with a body made up from a head joined to a segmented *thorax and *abdomen. Examples include shrimps, crabs, water fleas, lobsters, and woodlice.

cryolite A *mineral consisting of sodium aluminofluoride (Na_3AlF_6). It is used to lower the melting-point of aluminium oxide during the production of *aluminium by *electrolysis.

crystal A solid with a regular shape, containing *particles arranged in a *lattice. Crystals are formed by evaporating a solution or by cooling a molten substance until it solidifies. Crystals of a given substance are all the same shape but may have different sizes.   🖉

crystallization A *physical change where solid *crystals grow in a *saturated solution. The solution will crystallize as *solvent evaporates or the temperature is lowered.   🖉

current   see *electric current

cutting Part of a leaf or stem of a plant that has been cut off and put into soil. The part grows roots and becomes a new plant. This process is a type of artificial *vegetative reproduction.

Crystallize

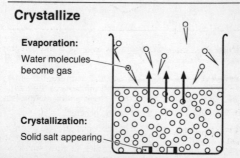

Evaporation:

Water molecules become gas

Crystallization:

Solid salt appearing

cyclone An area of low-pressure air covering part of the Earth. The pressure is lowest at the centre. Winds circle around a cyclone and towards the centre. Rain usually falls.

cylinder The circular sleeve inside which a *piston fits and slides.

cytoplasm Jelly-like material contained in a living *cell. It includes all the contents of the cell except the nucleus. Cytoplasm contains *organic compounds and *organelles.

D

Darwinism A theory of *evolution put forward by Darwin in 1859. It suggests that *mutation causes changes in the *characteristics inherited by organisms. Natural *selection forms new species from those organisms that have improved characteristics and that survive best in the environment.

data Information that has been collected together. *Computers use data together with a *program (both expressed in *binary code) to solve problems.

database Information about a particular subject, stored in a computer. E.g. television Ceefax is a news database. Information can be retrieved easily from a database.

DC Abbreviation for direct current. (see *electric current)

decay 1. Another term for *rot, applied to dead plants and animals as they *decompose. 2. The *nucleus of a *radioactive atom breaking into parts. *Alpha, *beta, or *gamma radiation is given off when a nucleus decays.

decibel (symbol dB) A unit that measures the difference between two power levels, usually sound or electrical signals. Where P_1 and P_2 are two different power levels, the difference (in dB) is calculated by the expression $10 \log_{10} P_1/P_2$.

deciduous Describing a *perennial plant that loses all its leaves during a certain time of the year. This helps the plant to conserve water while it is dormant and not growing. Roses and oak trees are examples of deciduous plants.

decompose

74

decompose To break down into separate smaller and simpler parts. Chemical *compounds decompose when they are heated or when electricity passes through them (see *electrolysis). *Saprophytes and other *decomposers decompose dead plants and animals.

decomposer An organism that causes the *decay of materials from dead plants and animals. *Bacteria, *fungi, and some *protozoa are decomposers. They play an important part in the *nitrogen and *carbon cycles by providing *nutrients for growing plants.

deficiency disease A disease that is the result of *malnutrition and caused by the lack of a *vitamin or *mineral or *amino acid in the diet. Examples are scurvy (lack of vitamin C) and kwashiorkor (lack of protein).

degree (symbol °) 1. A unit used for measuring angles. There are 360° in a full circle. 2. A unit used for measuring *temperature. The common unit is the degree Celsius (°C); the *SI unit is the kelvin (K). Water boils at 100 °C and freezes at 0 °C. 1 °C is equal to 1 degree kelvin (1 K). Water boils at 373 K and freezes at 273 K. 0 K is *absolute zero.

dehydration 1. Removing water from a substance, usually by heating. 2. A *chemical reaction where a *compound loses water, as atoms of hydrogen and oxygen are removed in the ratio 2 : 1. E.g. carbon results from the dehydration of glucose $C_6H_{12}O_6$ by concentrated sulphuric acid (a *dehydrating agent*).

deliquescent Describing a *hygroscopic solid that absorbs so much water from the surrounding air that it eventually forms a concentrated solution. E.g. sodium hydroxide.

denature 1. To alter the structure of a *nucleic acid or *protein by the action of heat or chemicals, causing the loss of its biological properties. Boiling an

egg denatures the protein and causes yolk and white
to solidify. 2. To add methanol or other poisonous
substances to ethanol (*alcohol) to make it unfit for
drinking.

dendrite The branched part at the end of a *neurone (nerve cell). Impulses flow from one nerve to
another through the dendrites. (see *synapse)

density The mass of 1 cubic metre of a substance.
E.g. the density of water = 1000 kg/m³; of air = 1.5
kg/m³; of steel = 7800 kg/m³. For any volume of a
substance,

$$\text{density} = \frac{\text{mass}}{\text{volume}}.$$

dental formula A method of showing the number
and type of teeth in an animal (usually a mammal).
A bear has a total of 42 teeth. Its dental formula is
$\frac{3\ 1\ 4\ 2}{3\ 1\ 4\ 3}$. Each half of the upper jaw contains 3 *incisors, 1 *canine, 4 *premolars and 2 *molars. Each
half of the bottom jaw contains 1 extra molar.

   Rabbit: $\frac{2\ 0\ 3\ 3}{1\ 0\ 2\ 3}$.   Human: $\frac{2\ 1\ 2\ 3}{2\ 1\ 2\ 3}$.

dentine see *tooth

deposition The building up of a layer of a substance
on a surface. Deposition of a metal occurs at the
*cathode during *electroplating. Deposition of *sedi-
ments on the ocean floor builds up *strata that
become *sedimentary rocks.

dermis see *skin

detergent A substance added to water that helps it
dissolve grease. Part of a detergent molecule is
attracted to water and the other part is attracted to
grease. Soap is called a soapy detergent. Washing-up
liquid, shampoo, and most washing powders contain
non-soapy (synthetic) detergents. These do not form
scum with hard water.

Detergent

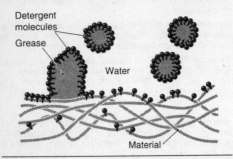

Detergent molecules

Grease

Water

Material

deuterium (heavy hydrogen) An *isotope of hydrogen, containing one neutron and one proton in its nucleus.

device An object made for a particular purpose. Examples include pulleys, electric motors, transistors, and hinges. (see *appliance)

dextrose see *glucose

diabetes A disease caused by *endocrine glands in the *pancreas not producing enough *insulin. The concentration of glucose in the blood becomes dangerously high. Diabetes can be controlled by a special diet, helped in some cases by regular injections of insulin.

diamond An *allotrope of *carbon. It is one of the hardest known substances and is used to make cutting tools. Cut and polished diamonds are used in jewellery because they appear to sparkle (by *refraction and *total internal reflection of light).

Diamond

A diamond is
a giant structure
of carbon atoms

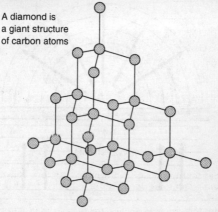

diaphragm 1. In mammals, a layer of muscles that
separates the *thorax (chest) from the *abdomen.
When the muscles contract, the diaphragm flattens
and moves downwards. This helps the lungs to fill
with air during *breathing. 2. A device used to control the amount of light entering a camera or microscope. A simple diaphragm is a thin sheet of metal or
plastic with a hole through its centre.

diastole see *blood pressure

diatomic Describing a molecule made up from two
atoms. Nitrogen N_2 and carbon monoxide CO are
diatomic gases.

Diffraction

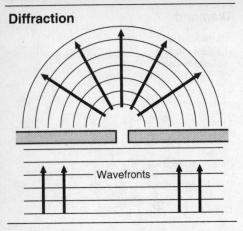

Wavefronts

diesel engine A type of *internal combustion engine that produces power in a similar way to a petrol engine, except that the fuel is not ignited by a spark. Fuel burns as it is sprayed into air that has been heated (to 500 °C) by compression. (see *compression ratio)

diesel oil A liquid containing *hydrocarbons with 16 to 20 carbon atoms per molecule. It is made from *crude oil and is used for heating and as a fuel in *diesel engines.

diet The different sorts of food and the amounts eaten by an animal. A balanced diet supplies the right amounts of *carbohydrate, *protein, *fats, *vitamins, minerals, water, and *dietary fibre.

dietary fibre (roughage) The part of food that is not
digested. *Cellulose makes up most of the dietary
fibre in the human *diet. It helps food to pass
steadily through the *alimentary canal.

dietary mineral see *essential element

diffraction The change in direction of a *wave as it
goes through a narrow gap. Diffraction scatters
*rays and bends *wavefronts, the effect being great-
est when the width of the gap is about the same as
the *wavelength.   ✍

diffusion The movement of *particles from a region
of high concentration to a region of lower concentration. Diffusion causes mixing. It is more rapid in
gases than liquids. E.g. two separate gases mix completely within minutes. Diffusion in solids is
extremely slow.   ✍

Diffusion

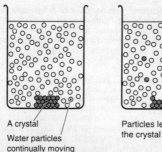

A crystal

Water particles
continually moving

Particles leave
the crystal

digestion In animals, breaking down food by *enzymes in the *alimentary canal. The result is substances the body can absorb and use. Digestion breaks down *proteins into *amino acids and *carbohydrates into *sugars. In humans, digestion takes place mainly in the *stomach and the *duodenum with the help of digestive enzymes and other chemicals. (see *saliva *amylase *pepsin *trypsin *bile)

digit A single number (numeral). There are ten digits in decimal counting: 0, 1, 2, 3, 4, 5, 6, 7, 8, 9. There are two digits in binary counting: 0, 1.

digital A measuring device that uses numbers. A digital watch uses numbers to indicate the time. A digital *computer uses numbers (*binary code) to solve problems. (see *analogue)

diode A device that allows an *electric current to flow one way only. It contains a piece of *semiconductor material joined to two connecting wires. Diodes are used as *rectifiers in electronic equipment (e.g. radios, television sets, computers).

diploid number The number of *chromosomes in each cell (excepting *gametes) of most organisms. The chromosomes are arranged in pairs, with one chromosome inherited from the female parent and one from the male parent. The diploid number for humans is 46 and the *haploid number is 23.

direct current (DC) see *electric current

discharge lamp A type of lamp that produces high-intensity light by passing an *electric current through a heated gas. Orange sodium street lights are a type of discharge lamp. *Xenon discharge lamps are used in photographic and disco flashlights and in lighthouses. They give out intense bluish-white light.

disinfectant A substance that kills or stops the growth of disease-causing *microbes but which will usually damage human *tissue. Disinfectants are used to clean kitchens, drains, and hospital equipment. A common example is bleach.

disk A thin circular sheet of magnetic material used to store *computer *programs or *data. Information can be recorded on to or read from a disk as it spins. A floppy disk is a flexible plastic disc coated with magnetic particles and contained in a stiff plastic envelope. A hard disk is rigid, is fixed permanently inside a computer, and can store much more information than a floppy disk.

dispersal The process that spreads *seeds away from their parent plant. Seeds are dispersed by wind, water, and animals.

dispersion The separation of white light into a *spectrum of colours (e.g. by a *prism or a raindrop). *Refraction bends the path of each colour in a ray of white light by an amount that depends on its *wavelength.   ✍

Dispersion

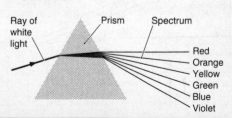

displacement reaction A type of *substitution reaction usually involving atoms or ions. E.g. iron reducing copper ions: $Fe(s)+Cu^{2+}(aq) \rightarrow Fe^{2+}(aq)+Cu(s)$; or chlorine oxidizing bromide ions: $Cl_2(aq)+2Br^-(aq) \rightarrow 2Cl^-(aq)+Br_2(aq)$.

dissociation The breakdown of a molecule or ion into smaller parts. Dissociation is usually a *reversible reaction, e.g. the dissociation of water $H_2O \Leftrightarrow H^+ + OH^-$.

dissolve To mix a solid or a gas with a liquid *solvent and form a *solution. When a substance dissolves, it is spread evenly throughout the solvent.

distillation A process used to separate a liquid from a solution. E.g. obtaining pure water from sea water. Pure vapour rises from the boiling liquid and is condensed to form pure liquid. Distillation also

Distillation

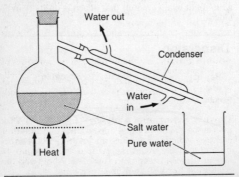

Water out

Condenser

Water in

Salt water

Pure water

Heat

separates liquids of widely differing *boiling-points. (see *condenser *fractional distillation) ✍

division A group of plants with similar *characteristics. The plant *kingdom is made up from several divisions. (see *classification)

DNA Deoxyribonucleic acid, a *nucleic acid found in the *chromosomes that make up the *genes in the nucleus of a living cell. Two strands of DNA pair together to form a double-helix molecule shaped like a twisted ladder. DNA controls the manufacture of *proteins and so controls the way an organism develops.

dominant allele An *allele that will override the effect of a *recessive allele in a pair. (NB each *characteristic of an organism is controlled by one or more pairs of alleles.) The alleles for brown eyes and dark hair are dominant. (see *heterozygous)

Doppler effect The apparent change in the *frequency of a *wave from a moving source heard or seen by a stationary observer. E.g. the frequency of sound of an car engine is raised as the car approaches an observer and is lowered as it moves away.

double glazing A window made from two panes of glass separated by a narrow air gap. The air sandwiched between the panes is a poor *conductor of *heat. Double glazing reduces the heat lost through the windows of a building.

Down's syndrome A disease in humans that affects mental ability and appearance. A *mutation causes a woman to produce an *ovum with 24 chromosomes. If the ovum is fertilized the baby has 47 chromosomes instead of the usual 46. About 1 in 700 babies has Down's syndrome.

drag The force acting against the motion of an object moving through a fluid. Drag is caused by *friction between the outside of the moving object and the fluid surrounding it. Drag increases with speed. It is lower for *streamlined objects and is greater in a liquid than in a gas.

drug A substance that alters the normal working of a person's mind or body. Drugs can be beneficial or harmful, legal or illegal. They include medicines, alcohol in drinks, nicotine in tobacco, cocaine, heroin, and caffeine in tea and coffee.

dry ice A solid made by cooling carbon dioxide gas. Dry ice *sublimes at −78 °C and is used as a simple and convenient refrigerant to keep things cold.

duct see *exocrine gland

ductility The ability of a metal to be pulled into a new shape without breaking. Copper is so ductile that that it can be drawn out (stretched) to make wire.

ductless gland see *endocrine gland

duodenum A part of the small *intestine in vertebrates. Food passes from the stomach into the duodenum where digestion is completed. Glands in the wall of the duodenum make digestive enzymes. *Bile from the *gall bladder and *enzymes from the *pancreas flow into the duodenum. (see *alimentary canal)

dynamo An electrical *generator, usually producing a direct *electric current.

E

E-number A code number that refers to a *food additive approved for use in the European Community (EC). An additive with an E-number is considered to be safe to use. The E-numbers and names of additives used in any processed or manufactured food must be printed on the wrapper.

ear In vertebrates, the sense organ that detects sound and controls balance. In humans and other mammals, sound-waves enter the outer ear (pinna) and vibrate the tympanum (ear-drum). The Eustachian tube connects the middle ear to the throat and keeps the air pressure equal on both sides of the tympanum. The three ossicles (malleus, incus, stapes) are bones that carry the vibrations from the tympanum and pass them through the middle ear and into the fluid-filled inner ear. Hairs inside the coiled cochlea vibrate and cause the auditory nerve to send impulses to the brain. The brain interprets these impulses as sound. Balance is controlled by the semicircular canals, which are three fluid-filled tubes at right angles to each other. When the head moves, the fluid moves and affects sensitive hairs connected to nerves. The brain uses information from these nerves to control muscles that keep the whole body balanced.   ✍

ear-drum see *ear

earth wire A safety wire connected to the metal case of an electrical appliance (e.g. a cooker). If a fault causes the live wire to touch the case, a large current flows through the earth wire and the fuse blows. Without the earth wire, a person could

Ear

Structure of the mammalian ear

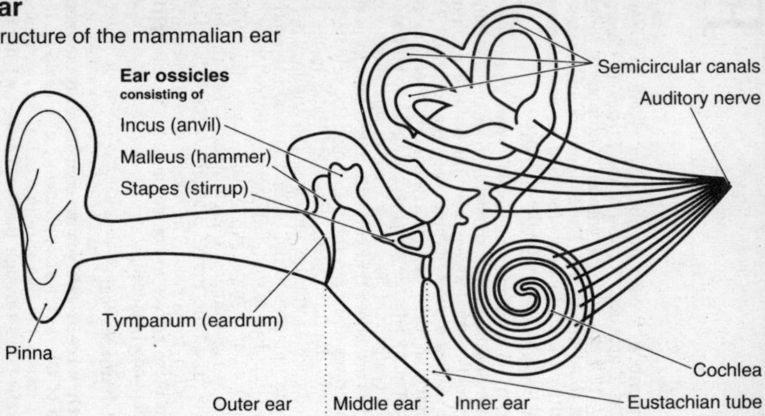

Ear ossicles
consisting of

Incus (anvil)
Malleus (hammer)
Stapes (stirrup)

Semicircular canals

Auditory nerve

Cochlea

Eustachian tube

Inner ear

Middle ear

Outer ear

Tympanum (eardrum)

Pinna

receive an electric shock from the case. Earth wires used to be connected to a metal rod buried in the ground. They are now usually connected to the neutral wire and to all the fixed metal objects (e.g. pipes, sinks) in the house.

echo The reflection of a *wave from a surface. A sound echo is a weaker repeat heard at the source after the original sound. Echoes of television signals from aircraft can cause 'ghosts' to appear on a TV screen.

eclipse An eclipse of the Sun happens when the Moon comes betweeen the Earth and the Sun. An eclipse of the Moon happens when the Earth comes between the Sun and the Moon. Planets can eclipse the light from stars.

ecology The study of how all living things depend on each other and the *habitats they live in. (see *ecosystem *environment *community)

ecosystem A *community of organisms together with the *environment they live in. An ecosystem on the land contains green plants (*producers) which are eaten by *herbivores (primary *consumers). The herbivores are eaten by *carnivores (secondary consumers). *Decomposers break down dead organisms and return *nutrients to the soil. The largest ecosystem is the biosphere. This is the whole world and all the organisms in it.

edaphic factor Part of the *environment that affects the life-style of an organism living in a *habitat. Edaphic factors include the composition and structure of the *soil and the type of plants that can grow in it.

effector A *muscle or *gland that is controlled by impulses from motor *neurones.

Eclipse

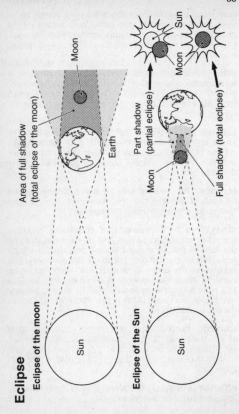

Eclipse of the moon

Moon

Area of full shadow
(total eclipse of the moon)

Earth

Sun

Eclipse of the Sun

Sun

Moon

Part shadow
(partial eclipse)

Full shadow (total eclipse)

Sun

Moon

Moon

Sun

effervescence Gas bubbles forming in a liquid as the result of a *chemical reaction. E.g. effervescence of carbon dioxide gas results when pieces of calcium carbonate are added to dilute hydrochloric acid.

efficiency A measure of how well a *machine (or other energy-changing device) changes *energy into useful work.

$$\text{Efficiency} = \frac{\text{work output}}{\text{energy input}} \times 100\%.$$

E.g. a petrol engine is 15% efficient. Every 100 *joules (J) of energy contained in the fuel are changed into 15 J of work and 85 J of waste heat and noise. Also,

$$\text{efficiency} = \frac{\text{power output}}{\text{power input}} \times 100\%.$$

E.g. a light bulb is 30% efficient. If the electrical power input is 100 watts, light output is 30 watts with 70 watts of waste heat.

effort The *force given to a *machine to move a load. (e.g. see *lever)

egg Another word for *ovum. It also refers more particularly to an ovum that has come out from the body of an egg-laying animal. The eggs of birds, reptiles, and insects contain a fertilized ovum and a yolk inside a protective shell or membrane. The eggs of fish and amphibia are jelly-like to allow external *fertilization.

egg cell see *ovum

elasticity The ability of a material that has been squashed or stretched by a force to return to its original shape when the force is removed. Rubber has high elasticity. Steel has greater elasticity than copper.

electrical energy The *energy of an electric charge flowing as a current around a circuit. E.g. a 6-

Electrical energy

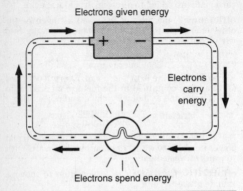

Electrons given energy

Electrons carry energy

Electrons spend energy

volt *battery gives 6 *joules of electrical energy to every *coulomb of charge it pushes around the circuit.   ✍

electric current A flow of electrical *charge. The electric current in a *circuit containing a *battery and bulb is the result of a steady flow of *electrons. A convention from before the discovery of the electron says that the current flows around the circuit from the positive (+) to the negative (−) connection on the battery. We now know that electrons actually flow from − to +. A direct current (DC) is a current flowing in one direction only. An alternating current (AC) continuously and regularly changes its direction and magnitude. The electric current in a liquid *electrolyte is the result of a flow of *ions. Electric current is measured in *amperes (amps).   ✍◻

electricity Effects that result from the movement of charged particles, usually *electrons. If charged particles collect and stay in one place, the effects are called *static electricity. If the particles flow steadily, the effects are those of an *electric current.

electric motor A *machine that converts electrical *energy into *mechanical energy. A simple DC (see *electric current) electric motor consists of a moving *rotor and a fixed *stator. The rotor consists of a coil of wire, which is an *electromagnet, spinning in a magnetic *field provided by the stator. An electric current enters and leaves the coil through carbon *brushes and a *commutator. Magnetic forces of attraction and repulsion between coil and stator turn the rotor. Every half turn, the commutator reverses

Electric current

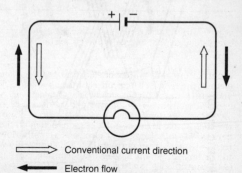

⇨ Conventional current direction

⬅ Electron flow

Electric motor

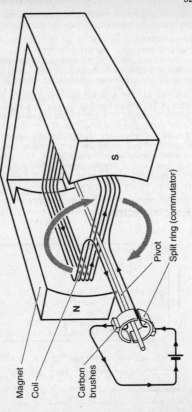

Magnet

Coil

Carbon
brushes

Pivot

Split ring (commutator)

S

N

the direction of the current in the coil, ensuring that the rotor continues to rotate in one direction only.
✍

electrochemical series A list of *elements arranged in order of their ability to lose electrons to their *ions in a solution. Elements that lose electrons more easily than hydrogen are electropositive. Elements that gain electrons more easily than hydrogen are electronegative. High electropositivity (e.g. potassium) and high electronegativity (e.g. fluorine) indicate high chemical reactivity.

electrode An electrical *conductor that gives out or takes in *electrons. *Cells, *transistors, and *valves have electrodes. (see *electrolysis)

electrolysis A *chemical reaction caused by passing an *electric current through an *electrolyte. The current enters and leaves the liquid electrolyte through *electrodes. Chemical reactions happen to the electrolyte (e.g. electrolysis of sodium chloride solution) or to the electrodes (e.g. purification of copper by electrolysis).
✍

electrolyte A *solution or a liquid that conducts an electric current. Molten *salts, and *aqueous solutions of *acids, *bases, and salts, are electrolytes. (see *electrolysis *cell)

electromagnet A coil of wire wound on an iron rod. A *magnetic field surrounds the rod for as long as an *electric current passes through the coil. Increasing both the number of turns of wire and the current increases the intensity (strength) of the magnetic field.
✍

electromagnetic energy *Energy contained in *electromagnetic radiation such as radio waves, light, X-rays, and gamma rays. The higher the *frequency (and shorter the *wavelength) of the radiation, the greater is its energy.

Electrolysis

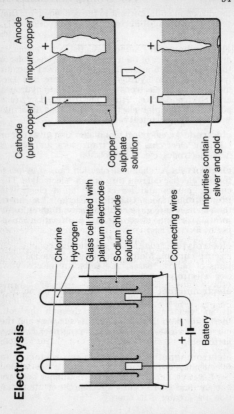

Anode (impure copper)

Cathode (pure copper)

Copper sulphate solution

Impurities contain silver and gold

Chlorine

Hydrogen

Glass cell fitted with platinum electrodes

Sodium chloride solution

Connecting wires

Battery

Electromagnet

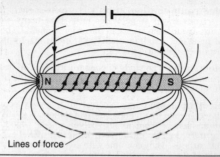

Lines of force

electromagnetic radiation *Energy that moves
as a transverse *wave at a *velocity of about 3×10⁸
m/s. The energy is contained in electric and magnetic *fields. The energy oscillates between the two
fields as the wave moves. The frequency of oscillation is the frequency of the wave. Electromagnetic
radiation includes *radio waves, *light, *X-rays, and
*gamma rays.

electromagnetic spectrum The range of all forms
of *electromagnetic radiation, listed in order of *frequency or *wavelength.

electromotive force (EMF) A measure of the
amount of energy given to each *coulomb of charge
provided by an electrical source (e.g. *battery, *generator). EMF is measured in *volts.

electron A particle contained in all *atoms and
*ions (except the hydrogen ion H⁺). It has a negative
*charge equal in value to the positive charge on a

*proton. The mass of an electron is about 1/1840 of the mass of a proton. (see *shell)

electronic configuration The arrangement of the electrons in an atom. It describes how many electrons are in each *shell, in the order K shell, L shell, M shell, etc. A sodium atom has 11 electrons. Its electronic configuration is 2,8,1.

electronics The study of communications and control and computing equipment that work by the flow of *electric currents in circuits.

electroplating Using *electrolysis to give a decorative or protective metal coat to an object. E.g. silver atoms ionize from a pure silver anode: $Ag(s) \rightarrow Ag^+(aq)+e^-$. The ions are attracted to a nickel jug at the cathode. They combine with electrons and form a coat of silver on the jug: $Ag^+(aq)+e^- \rightarrow Ag(s)$.   ✍

Electroplating

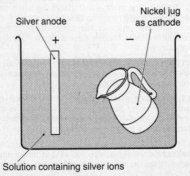

Silver anode

Nickel jug as cathode

+

−

Solution containing silver ions

electrostatic Describing any effect that is caused by *static electricity.

electrovalent bond see *ionic bond

element A substance that cannot be broken down into simpler substances. There are 92 naturally occurring elements, including *metals, *non-metals, and *metalloids. All the atoms in an element have the same number of protons and electrons. (see *isotopes)

embryo 1. Part of a *seed that grows into the shoot and root of a seedling. 2. A young animal growing inside its mother or inside an egg. In mammals, the embryo is the earliest stage of growth before it becomes a *foetus. An embryo does not have the same overall shape and parts as the adult.   ✍

EMF Abbreviation for *electromotive force.

emulsion A *colloid made from two liquids that do not mix. E.g. oil shaken with water forms an emulsion.

enamel see *tooth

endocrine gland (ductless gland) A *gland that makes a *hormone and releases it straight into the blood *circulation of an animal. Endocrine glands in humans include the *thyroid, *pituitary, and *adrenal glands, as well as the *testes, *ovaries, and part of the *pancreas.   ✍

endoskeleton see *skeleton

endosperm A part of some *seeds. Food stored in the endosperm helps the embryo to grow during *germination. (see *cotyledon)

endothermic Describing a chemical reaction that takes in *heat energy as the *reactants change into the *products. The result is a fall in temperature. E.g.

Embryo

A 28-day-old human embryo

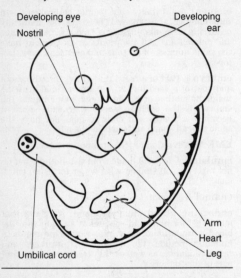

Developing eye

Nostril

Developing ear

Arm

Heart

Leg

Umbilical cord

roasting *limestone to make *lime: CaCO₃(s) → CaO(s) + CO₂(g). (see *energy level diagram)

energy The ability to do *work. The *SI unit of energy is the *joule (J). Energy can be converted from one kind into another. E.g. a light bulb changes *electrical energy into *light energy and *heat energy. (see *thermal energy *chemical energy

Endocrine gland

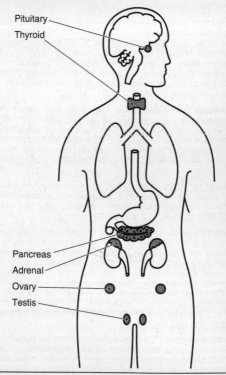

Pituitary

Thyroid

Pancreas

Adrenal

Ovary

Testis

*kinetic energy *potential energy *mechanical energy *electromagnetic energy *electrical energy *nuclear energy)

energy level The sum of the *potential and *kinetic energy that an electron has, depending on the *shell it is in. E.g. electrons in the L shell have higher energy levels than those in the K shell. When an atom takes in energy, an electron moves into a higher energy level. An atom gives out energy (*electromagnetic radiation) when an electron moves to a lower energy level.

energy level diagram A diagram that shows the changes in energy during a *chemical reaction. *Reactants take in energy that splits them into atoms or *ions (stage 1). Energy is given out as the atoms or ions combine to make the *products (stage 2). Reactions can be *exothermic or *endothermic overall.

↩

engine A *machine that changes the *chemical energy in a *fuel into *mechanical energy. (see *internal combustion engine *steam engine *turbojet engine *rocket engine)

entropy A measurement of the disorder of the particles in something. When energy changes from one form to another (e.g. when high-pressure steam drives a turbine), its source (the steam) becomes more disordered, and entropy rises. A state is reached (when the steam is completely expanded) where no more energy can be taken from the source. Entropy is now at a maximum. The total entropy of the Universe is steadily (and irreversibly) increasing.

environment Everything that affects an organism living in a *habitat. The complete environment of an organism includes the *biotic, *abiotic, *climatic, and *edaphic factors.

Energy level diagram

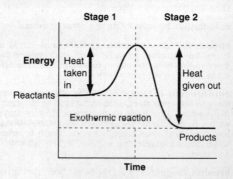

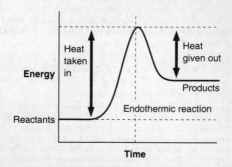

enzyme A *catalyst made up from *protein and contained in all living cells. Enzymes are involved in almost all the chemical changes in the *metabolism of cells. Each type of enzyme helps a particular chemical change to happen.

epidermis The outer *skin of a plant or animal.

epididymis In mammals, reptiles, and birds, a coiled tube attached to each *testis. It stores sperms made in the testis and passes them on to the *vas deferens. (see *sexual reproduction)

equilibrium 1. A state of balance where an object stays in the same place. Something is in equilibrium if a vertical line from its *centre of gravity stays within its base. Whether an object is in stable, unstable, or neutral equilibrium depends on how it behaves when moved slightly by a force. 2. see *chemical equilibrium   ✍

erosion The natural wearing away (*weathering) of the surface of the Earth and the movement of the resulting debris to other places. Erosion changes the shape of the Earth's surface. It is caused by wind, moving water in rivers and seas, and ice moving in glaciers.

erythrocyte A red *blood cell. Erythrocytes do not have a nucleus. They are shaped like flat discs and contain *haemoglobin which carries oxygen to the cells in the organism.

essential element (dietary mineral) Any of the metals and *salts that animals must take in with their food to remain healthy. Humans require tiny amounts of cobalt, zinc, copper, and iron (also known as trace elements) to make some types of *proteins and *enzymes. Calcium is required in larger quantities to make bones and teeth.

Equilibrium

Neutral equilibrium

Left alone, the ball stays where it is. Moved, it stays in its new position

Centre of gravity

Stable equilibrium

Even if you tip the cone a little, the centre of gravity stays over the base

Unstable equilibrium

The cone is balanced but not for long

Centre of gravity

---

ester A type of *organic compound that consists of a *carboxylic acid joined to an alcohol. E.g. methyl ethanoate:

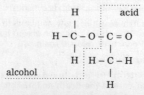

Many simple esters have strong fruity smells. They are used in industry as flavourings and solvents. Natural animal *fats and vegetable *oils are complex esters.

ethane A gas that is a *hydrocarbon and an *alkane. Its formula is C_2H_6. Ethane is found in *natural gas and dissolved in *crude oil.

ethanoic acid A liquid *carboxylic acid, formula CH_3COOH. It is present in vinegar and is used in industry to make important *raw materials and some plastics.

ethanol An *alcohol with the formula CH_3CH_2OH that results from the *fermentation of *sugars by *yeast. It is also made in industry from *ethene. Ethanol is present in alcoholic drinks and is also used as a solvent and as a fuel. (see *gasohol)

ethene A gas that is an *unsaturated *hydrocarbon and an *alkene. Its formula is C_2H_4. Ethene is made by *cracking hydrocarbons from *crude oil. It is used to make *polythene and important *raw materials such as *ethanol.

ethylene The old-fashioned name for *ethene.

ethyne A gas that is an *unsaturated *hydrocarbon and an *alkyne. Its formula is C_2H_2. Ethyne is made by adding water to solid calcium dicarbide. It burns in pure oxygen with a flame temperature of over 3000 °C, hot enough to weld and cut steel. Important *raw materials (e.g. *ethanoic acid) are made from ethyne.

etiolation The result of growing plants in dim light or darkness. They have pale (non-green) leaves, long shoots, and poorly developed roots.

Eustachian tube see *ear

eutrophication In a polluted river or lake, a process that causes the death and disappearance of most species of organisms. *Fertilizers or other nutrients drain into the water from the land. *Algae grow rapidly and then die. *Bacteria use dissolved oxygen from the water as they *decompose the algae. Other forms of life suffocate.

evaporate To change a liquid into a gas at a temperature below its *boiling-point. Applying a draught or increasing the temperature or the surface area of a liquid helps it to evaporate.

evergreen Describing a *perennial plant that bears leaves throughout the year. Some old leaves fall each year as new ones grow. Evergreen leaves are either small or waxy to cut down water loss and to withstand freezing conditions. Examples include holly, pine, spruce, and eucalyptus trees. (see *deciduous)

evolution A theory which attempts to explain the enormous variety of present-day living things, and how it happened. Evolution supposes that the *characteristics of a *species change by *mutation over many *generations. New characteristics sometimes improve the *adaptation of an organism to its environment. This organism will survive longer and produce more offspring which will inherit the improvement. The evolution of modern birds started over 200 million years ago from a type of lizard.

excretion Living organisms make waste substances that collect inside their *cells and *tissues. Waste substances include *carbon dioxide produced during *respiration, and *urea. Excretion removes these substances. Excretion in plants and simple animals is mainly by *diffusion. Complex animals excrete by using special *organs such as *lungs, *kidneys, and the sweat glands in *skin.

exhaust gases A mixture of gases continuously given out by an *engine as it burns *fuel and does work. The exhaust gases of vehicle engines contain

Excretion

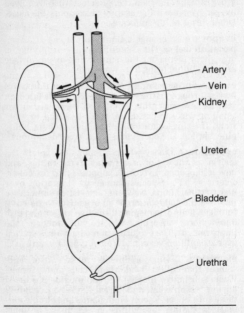

Artery

Vein

Kidney

Ureter

Bladder

Urethra

mostly carbon dioxide and water vapour. They also
contain smaller amounts of carbon monoxide, oxides
of nitrogen and sulphur, and unburned hydrocarbons that cause air *pollution. (see *tetraethyl lead)

exocrine gland A *gland that makes a liquid which then flows through a tube or opening called a duct. The duct leads on to the body surface or into a body cavity such as the *gut. Exocrine glands include part of the *pancreas, sweat glands, and *mammary glands.

exoskeleton see *skeleton

exothermic Describing a *chemical reaction that gives out heat as the *reactants change into the *products. The result is a rise in temperature. E.g. burning carbon in air: $C(s)+O_2(g) \rightarrow CO_2(g)$. (see *energy level diagram)

experiment An investigation to test a *hypothesis, *theory, or *law, or to discover facts (e.g. how substances behave, how animals react).

expression The result where an *allele causes an organism to have a particular *characteristic. The allele is expressed in the characteristic.

extinction The state where the number of organisms in a species falls to zero. This happens when the *environment changes so that the organisms are unable to survive.

eye An *organ that is sensitive to light. The human eye is protected by the tough white sclerotic layer (sclera). The clear conjunctiva protects the cornea. The aqueous humour is a watery liquid that fills the front of the eye and the vitreous humour is a clear jelly that fills the back. The transparent cornea and lens focus light on to the retina. Ciliary muscles pull on the suspensory ligaments, which alter the shape of the *lens and adjust its *focal length. The amount of light that reaches the retina is controlled by the iris. This is a ring of coloured muscle that surrounds a hole called the pupil. The retina contains light-sensitive *rod cells and *cone cells which send impulses along the optic nerve to the brain. The yellow spot

Eye

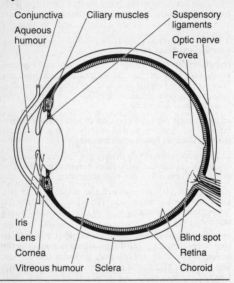

Conjunctiva · Ciliary muscles · Suspensory ligaments · Aqueous humour · Optic nerve · Fovea · Iris · Lens · Cornea · Blind spot · Vitreous humour · Sclera · Retina · Choroid

(*fovea) is the most sensitive part of the retina. The choroid is a black layer that stops light reflecting around inside the eye. The blind spot is the point where blood vessels and nerves join the eye. It is not sensitive to light.

eyepiece The *lens(es) in an optical instrument (e.g. microscope) closest to the eye and through which the *image is viewed.

F

F₁ (first filial) generation During an experiment in *genetics, the first *generation of offspring from selected parents.

factor Something that affects an organism living in a *habitat. (see *biotic factor *abiotic factor *climatic factor *edaphic factor)

faeces Solid waste matter from the *alimentary canal.

Fahrenheit scale A scale for measuring *temperature, now rarely used. Water boils at 212 degrees Fahrenheit (212 °F) and freezes at 32 °F (at *standard temperature and pressure).

fair test An *experiment that is carried out with all the known *variables held constant except two. E.g. a fair test of *Ohm's law varies the voltage across a conductor and measures the electric current through it, but keeps the temperature of the conductor constant. A change in temperature (or length or thickness) would affect the resistance and make the test unfair.

Fallopian tube (oviduct) In female mammals, the tube that carries egg cells (*ova) from the *ovary to the *uterus. *Fertilization takes place in the Fallopian tube when *sperms from a male meet an ovum. (see *sexual reproduction)

family A group of living organisms with similar *characteristics. Several families make up an *order. Lions are in a family called the cats, which includes tigers, leopards, jaguars, and domestic cats. (see *classification)

farad (symbol F) The *SI unit of electrical capacitance. A *capacitor has a capacitance of 1 F if it is charged with 1 *coulomb and has a *potential difference of 1 volt across its plates.

Faraday constant (symbol F) The amount of *charge carried by 1 *mole of *electrons, approximately 96 500 *coulombs.

fat A solid substance used by plants and animals as a store of food energy. Fats are usually *esters made up from *glycerol and *fatty acids. They have about twice the energy content of *carbohydrates. Fats also insulate animals against heat loss and form a protective cushion around organs.

fatty acid A substance contained in most *fats and natural *oils. Fatty acids are made up from a long straight *hydrocarbon molecule with a carboxyl —COOH group at one end. (see *carboxylic acid)

fault A break in the Earth's *crust where a very large mass of rock splits and the two parts slide past each other.

feedback The control of a process or system, using information from its output or effects. Feedback from a *thermostat switches a refrigerator on and off and controls its temperature. Biological feedback controls the numbers of each *species in a *food chain. Feedback in an electronic *amplifier involves routeing part of the output signal back to the input. There are two sorts of electronic feedback: negative feedback subtracts from the input signal and lowers the *gain; positive feedback adds to the input signal and raises the gain, with the result that the amplifier becomes an *oscillator.

femur The thigh-bone in humans and many other land vertebrates. (see *skeleton)

Fault

Original block

Normal fault Fault plane

Reverse fault

Strike-slip fault

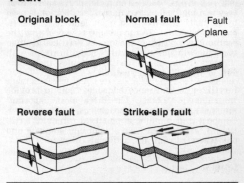

fermentation A process carried out by yeasts and some *bacteria. These organisms take in sugars and change them into ethanol (*alcohol) and carbon dioxide. Fermentation is a type of *anaerobic respiration and is used in baking and brewing.

fern A type of plant with large leaves that uncurl upwards as they grow. Ferns use *spores to reproduce. An example is bracken.

fertile 1. A living organism is fertile if it is able to reproduce. Human females are fertile between the ages of about 12 and 50 years. 2. Plants grow well in fertile *soil. It is made from the correct mixture of solid particles and contains suitable amounts of plant *nutrients.

fertilization The joining of male and female *gametes during *sexual reproduction. In animals, a male *sperm joins with a female *ovum. External fertilization is used by animals that reproduce in water (e.g. fish, frogs, toads). The male sprays sperm over eggs laid by the female. Internal fertilization is used by animals that reproduce on land (e.g. insects, birds, mammals). The male injects sperm into the reproductive organs of the female. Fertilization then takes place inside the female's body. In plants with *flowers, male pollen fertilizes ovules inside the female ovary. (see *pollination)

fertilizer A substance added to *soil to provide *nutrients for plants. The three most important nutrients are nitrogen, phosphorus, and potassium. Compost and manure are natural (organic) fertilizers. Chemicals such as ammonium sulphate and potassium nitrate are artificial (inorganic) fertilizers.

fetus see *foetus

fibre An object shaped in a long thin strand. Cotton and wool are natural fibres, as are *muscle fibres and *nerve fibres. Man-made fibres (e.g. nylon, polyester) are *polymers manufactured from chemicals made from *crude oil.

fibreglass *Fibres of glass spun into threads or woven into cloth. It is used to make fireproof clothing and *glass-reinforced plastic.

fibula The smaller of the two bones that join the ankle to the knee in humans and many other land vertebrates. (see *skeleton)

field The space in which one object exerts a force on another. A mass is surrounded by a gravitational field that attracts other masses. A *magnet is surrounded by a magnetic field that can attract or repel other magnets. A *charge is surrounded by an electric field that can attract or repel other charges. The magnitude (strength) of the field decreases with distance from the object.

field of vision How far around itself an animal can see without moving its head or eyes. The separate fields of vision of each eye make up the overall field of vision. Some animals (e.g. humans, owls) have binocular vision because their eyes point forwards. An object can be seen by both eyes at the same time, allowing the animal to estimate distance.

filament 1. see *flower. 2. A thin wire with a high electrical *resistance. A filament becomes hot when an *electric current passes through it. The filament in an electric light bulb glows white-hot and gives out light. The filament in an electronic *valve or *cathode ray tube gives off electrons.

filter 1. A transparent sheet of glass or plastic that removes colours from light that passes through it. E.g. white light is a mixture of red, green, and blue. A red filter removes blue and green from white light. 2. see *filtration

filter bed A layer of sand, gravel, and stones used to remove solids from water in a waterworks.

filter funnel see *filtration

filter paper see *filtration

filtrate The clear liquid that results when a *suspension is filtered.

filtration A process that separates solid particles from a *suspension. A filter paper contains microscopic holes which allow liquid but not solid particles to pass. The filter paper fits inside the filter funnel.

fire triangle A diagram that shows the three conditions needed for burning: air/oxygen, fuel, and heat. Removing any one of them will put out the fire.

fission see *binary fission *nuclear fission.

flagellum (pl. flagella) A long whip-like thread, longer than a *cilium. It is an outer part of some *unicellular organisms and *sperms. It beats and moves the organism through a liquid. Some *bacteria and *protozoa have groups of flagella.

flame test A simple test to identify some metals contained in *compounds. A platinum wire is dipped in concentrated hydrochloric acid and then in the compound, and is then held in a Bunsen flame. The colour of the flame indicates the following metals: green—barium; brown/red—calcium; crimson—lithium; lilac—potassium; yellow/orange—sodium; red—strontium.

flammable (inflammable) Describing a substance that easily catches fire, usually a gas or the vapour from a *volatile liquid. *Methane and petrol are highly flammable.

flask A container with a round body and a narrow neck, used for holding liquids.

flatworm A type of *worm which has a flat non-segmented body with a mouth at one end. Freshwater

Flask

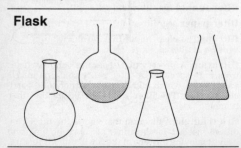

flatworms live in ponds and streams. Flukes and tapeworms live in the bodies of other animals and cause illness.

Fleming's left-hand rule A rule showing the direction of the *force acting on a conductor when the *electric current flowing in it is at right angles to a *magnetic field. It can be used to find the direction of rotation of simple *electric motors.   ✍

Fleming's right-hand rule In problems involving *induction, a rule showing the direction of flow of an *electric current in a conductor that is moving in a *magnetic field.   ✍

Fleming's Left-Hand rule

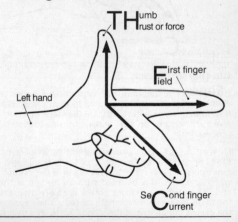

Fleming's Right-Hand rule

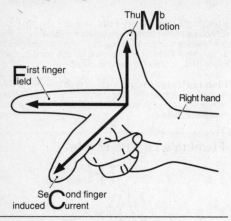

Thu**M**b Motion

First finger ield

Right hand

Se**C**ond finger induced **C**urrent

flip-flop see *bistable

flow chart A diagram that shows how a series of changes in a process are related to each other. E.g. flow charts are used to illustrate the steps in a computer *program or the stages in an industrial process.

flower A part of all plants that are *angiosperms. Flowers contain the *organs used for *sexual reproduction. The male reproductive organ is the stamen, which consists of a filament with an anther at the end. The anther produces pollen grains which are the male *gametes. The female reproductive organ is the carpel, consisting of the stigma, style, and ovary.

Flower

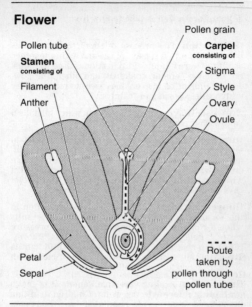

Pollen grain

Pollen tube

Stamen
consisting of

Filament

Anther

Carpel
consisting of

Stigma

Style

Ovary

Ovule

Petal

Sepal

- - - -
Route
taken by
pollen through
pollen tube

The ovary contains one or more ovules (female gametes). Each ovule contains an *ovum that will grow into a *seed after *fertilization. *Pollination transfers pollen from an anther to a stigma. A pollen tube grows from the pollen grain, down through the style and into the ovary. Fertilization occurs when the nucleus of the pollen cell travels down the pollen tube and joins with the nucleus of an ovule. ✍

fluid A *gas, a *liquid, a finely powdered solid, or any substance that is able to flow from one place to another.

fluidization The process of small solid particles being held up in a stream of gas and acting together as if they were a liquid. Fluidization is used in industry to move cement, coal-dust, and other fine powders. Fluidization also enables gases to react rapidly with a finely powdered *catalyst.

fluorescein A dye that dissolves in water to make a yellowish-green fluorescent solution.

fluorescence The ability of some substances to absorb light of one frequency and emit light of a different frequency. Fluorescence by optical brighteners in washing powder make clothes appear 'extra white and clean' by absorbing invisible *ultraviolet rays from sunlight and emitting white light.

fluorescent lamp A type of electric lamp, often in the form of a long tube. *Ultraviolet light results when a current flows through a heated low-pressure gas (e.g. neon). The ultraviolet light causes *fluorescence of a coating inside the tube, and white light is given out. Fluorescent lamps are more efficient than ordinary electric *light bulbs.

fluoridation Adding very small amounts of *salts containing fluorine to tap water. Children drinking fluoridated water grow teeth that have increased resistance to decay.

fluorine (symbol F) A pale yellow gaseous *nonmetal *element that is a member of group VII in the *periodic table (the *halogens). It exists as molecules F_2, and is poisonous and extremely reactive. Fluorine is extracted by *electrolysis of molten sodium fluoride and is used to make *organic fluorine compounds (e.g. *PTFE).

flywheel A heavy wheel that spins and stores energy. A flywheel has large *inertia due to its large *mass. (see *internal combustion engine)

FM Abbreviation for frequency modulation. (see *modulation)

focal length The distance between the centre of a lens and its principal focus. The more curved the surface of a lens, the smaller is its focal length. (see *convex lens *concave lens)

focal point Another name for principal focus. (see *convex lens *concave lens)

focus 1. To adjust an optical device (e.g. TV, microscope) to obtain a sharp *image or picture. 2. principal focus: see *convex lens *concave lens

foetus In mammals, a young animal growing inside its mother and with the same overall shape and parts as the adult. A human baby is called a foetus after it has been growing for eight weeks. Before this time it is called an *embryo.

fold A bend in a layer of *sedimentary rock caused by movements of the Earth's *crust. The two simplest types of fold are the arch-shaped anticline and the basin-shaped syncline. ✍

follicle A group of cells that surround and protect or nourish part of an animal. A hair follicle surrounds the root of a hair. Follicles in an *ovary surround growing *ova (eggs).

follicle-stimulating hormone (FSH) A *hormone made in the *pituitary gland of a mammal. It helps with the production of *ova (egg cells) in females and *sperms in males.

food additive A substance added to manufactured or processed food. Food additives are commonly used to improve the colour or keeping properties of

Fold

Monoclinal fold

Anticlinal fold

Synclinal fold

Overturned fold

Direction of
pressure →

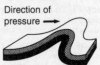

Recumbent fold

Direction of
pressure →

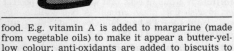

food. E.g. vitamin A is added to margarine (made from vegetable oils) to make it appear a butter-yellow colour; anti-oxidants are added to biscuits to delay spoilage by the air. (see *E-number)

food chain A list of organisms, starting with a (plant) *producer, showing how each organism depends on another for food. An example is: cabbage → caterpillar → blackbird → hawk. Energy flows through a food chain in the direction: producer →

primary *consumer (*herbivore) → secondary consumer (*carnivore) → tertiary consumer (larger carnivore). (see *food web)

food web Organisms grouped in a diagram that shows how each one eats a range of other different organisms. A food web is made up from a number of interconnected *food chains.   ✍

force 1. A push or a pull that causes an object to accelerate, slow down, or change shape. In the case of *acceleration, force = mass×acceleration. The unit of force is the *newton (N). 1 N is the force that gives 1 kg of mass an acceleration of exactly 1 m/s^2. 2. see *reaction

formic acid The old-fashioned name for *methanoic acid.

formula *Chemical symbols written to show the *elements in a *compound. The formula of an *ionic compound also shows the ratio of the elements. E.g. $CaCl_2$ means there are twice as many chlorine ions as calcium ions. The molecular formula of a *covalent compound shows the elements and the number of each in one *molecule. E.g. ethane, molecular formula C_2H_6, contains 2 carbon atoms and 6 hydrogen atoms in each molecule. The empirical formula shows the simplest ratio of the atoms, e.g. ethane, CH_3. The structural formula shows how the atoms are arranged, e.g. ethane, $CH_3.CH_3$. The geometrical formula shows the 3-dimensional shape of the molecule, e.g. ethane:

$$
\begin{array}{ccc}
\text{H} & & \text{H} \\
| & & | \\
\text{H} - \text{C} & - & \text{C} - \text{H} \\
| & & | \\
\text{H} & & \text{H}
\end{array}
$$

formula mass The mass of a *molecule or a group of *ions contained in a substance, on the scale where

Food web

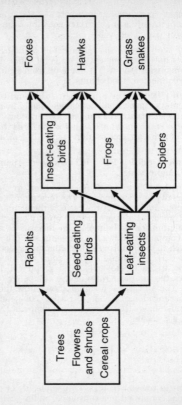

an atom of the *isotope carbon-12 (^{12}C) is fixed at exactly 12.0000. Formula mass has no units. It is calculated by adding the *relative atomic masses (r.m.m.) of the atoms, according to the written formula of the substance. E.g. carbon dioxide (CO_2) r.m.m. is 12+16+16 = 44; sodium chloride (Na^+Cl^-) r.m.m. is 23+35.5 = 58.5.

fossil fuel *Coal, *crude oil, *natural gas (*methane), peat, and *tar sands. Fossil fuels are the remains of organisms that became buried millions of years ago.

fovea (yellow spot) The most sensitive part of the retina in the human *eye. It is opposite the lens and contains a large number of *cone cells

f.p. Abbreviation for *freezing-point.

fraction A substance or group of substances with similar *boiling-points that are separated from a liquid mixture by *fractional distillation. E.g. diesel oil is the fraction collected from *crude oil at 220–350 °C.

fractional distillation (fractionation) A process that separates mixtures of liquids with boiling-points that are close together. A mixture of vapours from the boiling liquid ascends through the *fractionating column. The vapour of the substance with the lowest boiling-point emerges from the top of the column. The vapours of liquids with higher boiling-points condense and flow back down the column. (see *oil refinery)

fractionating column A long narrow vertical tube used in *fractional distillation. It contains trays or glass balls that cause the ascending vapour to mix thoroughly with descending liquid. The temperature is higher at the bottom than at the top. (see *oil refinery)

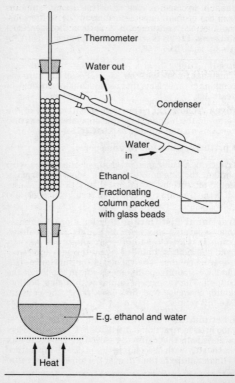

Fractional distillation

Thermometer

Water out

Condenser

Water in

Ethanol

Fractionating
column packed
with glass beads

E.g. ethanol and water

Heat

fractionation see *fractional distillation

Frasch process A method of extracting *sulphur
from the ground. Superheated steam at 180 °C melts
the sulphur (*melting-point 113 °C) and compressed
air forces it to the surface.   ✍

Frasch process

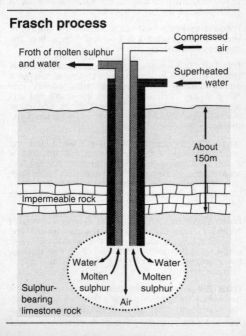

free fall The movement of an object attracted by *gravitational force and not affected by *drag or *buoyancy.

freeze To change a liquid into a solid by cooling.

freezing-point (f.p.) The temperature at which a liquid changes into a solid. For a pure substance, it is the same as the *melting-point.

frequency For a *wave or vibration, the number of oscillations completed in 1 second. The unit of frequency is the *hertz (Hz).

friction The force that acts against the movement of things that are in contact and sliding across each other. Friction converts *mechanical energy into *heat energy. Friction between the brake pad and wheel slows a bicycle. Oil in engines and other machines reduces friction between moving parts and increases their *efficiency.

front The dividing-line where masses of cold and warm air meet. Cold air moving into an area of warm air forms a cold front. Warm air moving into an area of cold air forms a warm front. Clouds form in both cases as the warm air rises and cools. Rain usually falls from the layered clouds associated with a warm front.

froth flotation A process that separates a wanted *mineral from the unwanted *gangue in an *ore. Powdered ore is added to a tank containing water and a special detergent. Bubbles of air rise through the mixture and carry the mineral to the surface. The gangue sinks.

fruit The part of a *plant that contains and protects its *seeds. The outside part of a fruit is made up from the ovary wall of the plant. Some fruits are dry (e.g. walnuts) and some are fleshy (e.g. tomato).

FSH Abbreviation for *follicle-stimulating hormone.

fuel 1. A substance that burns and can be used as a source of heat energy. E.g. *fossil fuels, wood, *biofuels. 2. A nuclear fuel is a *radioactive substance (e.g. *uranium, *plutonium) used inside a *nuclear reactor.

fuel oil A *viscous liquid containing *hydrocarbons with 30 to 40 carbon atoms per molecule. It is made from *crude oil and is used as a fuel in ships and *power-stations.

fulcrum The point on which a *lever rests and about which it turns.

fungicide A substance that kills *fungi. Some fungi are *parasites that cause damage and disease to plants and animals. Farmers spray crops with fungicides to prevent damage by mildew. Some ointments contain fungicides that cure diseases like athlete's foot.

fungus (pl. fungi) A type of simple land organism that lacks *chlorophyll and is a member of the fungi *kingdom. Moulds, mildews, mushrooms, and toadstools are all fungi. They exist as a tangled mass of threads called *hyphae and feed from other organisms. They are *parasites or *saprophytes. Fungi use *spores to reproduce.

fuse 1. A thin wire included in some electric circuits to protect against faults caused by a *short circuit. If a dangerously high current flows in the circuit the fuse becomes hot because it has a high *resistance. The wire in the fuse melts, breaking the current flow and switching off the circuit. 2. Another word for *melt.

fusion 1. Another word for melting. 2. see *nuclear fusion

G

gain The output voltage of an *amplifier compared to its input voltage.

$$\text{Voltage gain} = \frac{\text{output voltage}}{\text{input voltage}}.$$

galaxy A vast collection of stars, gas, and dust, usually in the shape of a thick disc. The Sun belongs to a galaxy (called the Milky Way or The Galaxy) that contains about 10^{11} stars and is 100000 light-years in diameter. The nearest galaxy to our own is 2 million light-years away. There are many millions of galaxies in the *Universe.

galena An *ore used in the production of *lead. It contains lead sulphide PbS.

gall bladder An *organ that stores *bile made in the *liver of vertebrates. The gall bladder lies under the liver and passes bile into the *duodenum. (see *alimentary canal)

galvanize To rustproof *steel by coating it with a layer of zinc. Steel is galvanized by dipping in a tank of molten zinc.

galvanometer A very sensitive type of *ammeter, used to detect and measure very small *electric currents flowing in a conductor.

gamete A reproductive cell in an organism that reproduces sexually. Examples are *sperm cells and *ova in animals, and pollen and *ovules in *flowers. Gametes have half the number of *chromosomes compared to other cells in the organism. A female gamete joins with a male gamete during *fertilization to form a *zygote.

gamma radiation *Electromagnetic radiation emitted by the nuclei of some *radioisotopes. It is so highly penetrating that 25 mm of lead only halves its strength. (see *nuclear radiation)

gangue The part of an *ore that does not contain the mineral required for metal extraction. A rich iron ore may contain 40% gangue; a poor copper ore may contain 99% gangue.

gas A *state of matter where a substance is a *fluid that completely and evenly fills its container. The *particles in a gas are widely spaced and move very rapidly, colliding with each other and the walls of the container. These collisions produce the pressure that the gas exerts.

gas exchange In *lungs and *gills, the absorption of oxygen by the blood and the removal of carbon dioxide from the blood.

gas laws Laws that describe how the volume (V) of a gas changes when temperature (T) or pressure (P) are varied. If 1 is the initial state and 2 the final state,

$$\frac{P_1 \times V_1}{T_1} = \frac{P_2 \times V_2}{T_2}.$$

gasohol A liquid fuel that is a mixture of petrol and alcohol. The alcohol is made by fermenting sugarcane or grain. Sunny countries like Brazil can make alcohol that is cheaper than imported petrol. Gasohol usually powers cars.

gasoline see *petrol

gas syringe A piece of apparatus used to measure the volume of a sample of gas. It consists of a transparent cylinder with a scale marked in cm^3 and a freely moving *piston (plunger).

Gas syringe

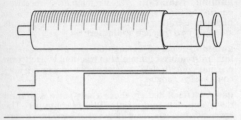

gate see *logic gate

gear A wheel that has ridges called cogs (teeth) sticking out around its circumference. Two gears of different diameter turning together are a *machine that alters the speed of rotation or the turning effort (*torque) applied to it. ✍

Geiger counter A device for detecting *nuclear radiation. The radiation ionizes a gas in a tube. The *ions and free *electrons that result are attracted to charged *electrodes. A pulse of current flows through an external circuit and works a counter.

gel A *colloid that has set into a solid form. Examples include table jelly, silica gel, and hair-styling gel.

gene A part of a *chromosome inside the cells of a living organism. Genes are made up from *DNA coated with *protein. Pairs of genes control each of the *characteristics of an organism. E.g. there are genes controlling eye colour and other genes controlling hair colour. The genes for a given character-

Gear

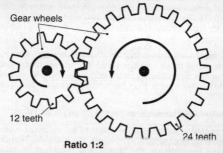

Gear wheels

12 teeth

24 teeth

Ratio 1:2

istic are always located at the same place in a chromosome. (see *allele *phenotype *genotype)

gene pool All the *genes and their different *alleles that are present in a *population of a species.

generation The members of a family or group of animals that have similar ages. Parents form one generation and their children are the next generation.

generator A *machine that converts *mechanical energy into electrical energy. A simple DC generator has the same arrangement of parts as a DC *electric motor. It generates pulses of *electric current that flow in one direction only. An alternator generates an alternating electric current (AC). It has a *commutator consisting of two separate rings (slip rings), one connected to each end of the *rotor coil. A carbon brush presses against each ring. (AC, DC—see *electric current)

genetic code (base code) A set of instructions contained within the structure of *DNA which controls the manufacture of proteins. Protein manufacture affects the function of each cell and so the genetic code influences the *characteristics of the whole organism.

genetic engineering In *biotechnology, altering the genes in a microbe so that it produces useful substances. Bacteria that normally produce gas can now make *hormones (e.g. *insulin), vaccines, and vitamins.

genetics The study of *heredity and *variation.

genitals (genitalia) The parts of an animal's reproductive organs that show on the outside. Human male genitals are the *penis and *scrotum. Human female genitals are the two folds of flesh that make up the *vulva.

genotype A set of *genes on the *chromosomes of an organism that produce a *phenotype. Genotype also means the complete set of genes contained in one cell of an organism. (see *allele)

genus (pl. genera) A group of living organisms with very similar *characteristics. Several genuses make up a *family. Lions are in a genus called the large cats, which includes tigers and leopards. (see *classification)

geology The study of the origin, structure, and composition of the Earth.

geostationary satellite A type of *artificial satellite that stays in a fixed position 35900 km above a point on the Earth's equator. The *period of its *orbit is the same as the period of the Earth's rotation (1 day). Three geostationary satellites give world-wide communications coverage.

geothermal energy Heat energy that comes from hot rock deep under the Earth's surface. Boreholes carry water down to the hot rock. The water boils and steam returns to the surface through other boreholes. The steam powers *turbines that drive electric *generators. The hot water that results heats homes and factories. (see *renewable energy)

germination The growth of a new plant (seedling) from a *seed or *spore. A seed needs water, warmth, and oxygen for germination to happen. Water causes the seed to soften and swell. The embryo grows with the help of food from either the *cotyledons or the *endosperm. In the embryo, the *plumule becomes the main *stem of the seedling and the *radicle becomes the main *root.

gestation In animals, the length of time between the *fertilization of an *ovum and the birth of offspring. In humans, gestation is called pregnancy and lasts about 40 weeks. Gestation lasts 3 weeks in a mouse and 96 weeks in an elephant.

giant structure (macro-structure) A type of substance that contains *particles arranged in a regular repeating pattern. *Diamond is a giant structure of carbon atoms. A *salt (e.g. sodium chloride) *crystal is a giant structure of ions. A sugar crystal is a giant structure of sucrose molecules. (see *lattice)

gill 1. The organ used by animals which live in water to obtain oxygen and give out carbon dioxide. Gills are parallel sets of thin *membranes that contain blood *capillaries. Water enters the mouth and flows past the membranes, exchanging oxygen and carbon dioxide with the blood. The water flows out through openings called gill slits. Some other aquatic animals (e.g. sea slugs) have external gills. 2. One of the ridges that make *spores on the underside of a mushroom.

Giant structure

Sodium chloride – a giant structure of ions

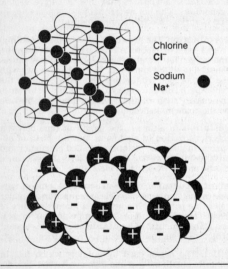

Chlorine
Cl⁻

Sodium
Na⁺

---

gland An organ in a plant or animal that makes a particular substance. There are two types of gland in mammals: *exocrine glands and *endocrine glands. Substances made by glands include *hormones, *enzymes used in *digestion, sweat, and *saliva.

glass A transparent solid made by heating together *lime, sodium carbonate, and sand. Glass is noncrystalline, which means that its atoms are in a random arrangement. One result of this is that glass does not have a definite melting-point, instead becoming steadily softer as its temperature increases.

glass-reinforced plastic (GRP) A solid *composite material that is commonly called fibreglass, made by mixing two resins with reinforcing *fibres of glass. *Polymerization causes the mixture to solidify. GRP is used to make the hulls of small boats and electronic circuit boards.

glomerulus see *kidney

glucagon A *hormone made by *endocrine glands in the *pancreas. It helps to regulate the amount of *glucose in the blood. Glucagon has the opposite effect to *insulin. It causes the *liver to change *glycogen into *glucose.

glucose (dextrose) A *sugar that is present in all plants and animals. It is made in plants by *photosynthesis. Glucose is the result of *digestion by animals of *carbohydrates. Glucose is used by plants and animals to release energy during *respiration.

glycerine see *glycerol

glycerol (glycerine) Propane 1,2,3–triol, a colourless, sticky liquid that is soluble in water and tastes sweet. Many natural *fats and *oils are *esters containing glycerol combined with *fatty acids.

```
        H   H   H
        |   |   |
    H – C – C – C – H
        |   |   |
       OH  OH  OH
```

glycogen A type of *carbohydrate similar to *starch, used by animals to store energy in their muscle cells and *liver. It is made up from *glucose molecules joined together in clusters. (see *insulin)

gold (symbol Au) A soft yellow *transition metal *element that is very unreactive. Gold *ore contains particles of the free metal. It does not tarnish and is used for jewellery and dentistry and to coat parts of some electronic devices.

gonad An organ that produces *sperms or *ova (*gametes) in animals. *Ovaries and *testes are gonads. In many animals, gonads also produce *hormones.

gram (symbol g) A unit of *mass equal to one thousandth of a *kilogram.

granite A very hard crystalline *rock, usually containing *quartz crystals. Most types of granite are *igneous rocks, although some are *metamorphic.

graphite An *allotrope of *carbon and one of the few *non-metals that can conduct electricity. It is soft and slippery and is use to make pencil leads, *electrodes, and high-temperature lubricants. ✍

gravitational force The force of attraction between the masses of two objects. The greater the mass of each object and the closer together they are, the greater the gravitational force. *Weight is the effect of gravitational force acting on the mass of an object.

greenhouse effect The raising of the overall temperature of the world, caused by carbon dioxide in the atmosphere trapping *solar energy. The amount of carbon dioxide is increasing due to the burning of *fossil fuels and deforestation.

group A vertical column of elements in the *periodic table. All the elements in a group have similar properties, because they all have the same number of outer-shell electrons. The elements show a trend in

Graphite

Graphite is made of flat
sheets of carbon atoms

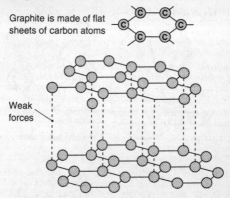

Weak
forces

properties down the group as the number of inner
*shells increases.

growth curve A graph that shows the growth of an
organism over time. A simple growth curve results
from plotting body mass against age. Spurts of
growth and difference between sexes are shown
more clearly when *rate* of growth is plotted against
age.

GRP Abbreviation for *glass-reinforced plastic.

guard cell A type of cell found in pairs especially
on the underneath surface of a *leaf. Each pair surrounds a hole called a *stoma. The shape of the guard
cells depends on the amount of water in the leaf. This

Guard cell

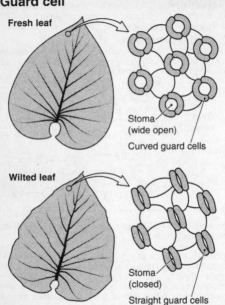

Fresh leaf

Stoma
(wide open)

Curved guard cells

Wilted leaf

Stoma
(closed)

Straight guard cells

controls the size of the stoma and the amount of
water vapour lost by the leaf.

gullet see *oesophagus

gut see *alimentary canal

gymnosperm A plant that produces seeds with the help of male and female cones. *Evergreen trees such as pines, firs, and spruces are gymnosperms. The wind carries *pollen from the male to the female cones.

gypsum A mineral containing *hydrated calcium sulphate $CaSO_4.2H_2O$. It is used in the manufacture of cement, plaster, paper, and rubber.

H

Haber process The industrial method used to make *ammonia. An iron catalyst combines nitrogen and hydrogen at 200 atmospheres pressure and 450 °C. Nitrogen is obtained from *liquid air by *fractional distillation and hydrogen is obtained from *methane by *steam reforming.

habitat A place where organisms live. Different habitats include heathland, grassland, woods, marshes, ponds, and the various parts of the sea. Each habitat contains its own *community of different plants and animals. (see *environment)

haematite A dense dark-red mineral containing *iron oxide Fe_2O_3. It is an important *ore used in the production of iron.

haemoglobin A chemical contained in *erythrocytes (red *blood cells). It enables erythrocytes to carry oxygen from the lungs to cells in the body *tissues.

haemophilia An inherited disease in humans where the blood does not clot properly. It is caused by a *recessive allele on the X *sex chromosome. A woman has two X chromosomes. An allele on one for haemophilia will be masked by a dominant allele for normal clotting on the other. The woman is a carrier for haemophilia. If her son inherits the haemophilia allele, it will not be masked by the Y sex chromosome that pairs with the X. He will have the disease and his daughters will be carriers.

Half life

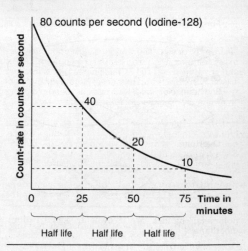

80 counts per second (Iodine-128)

Count-rate in counts per second

40

20

10

0   25   50   75   Time in
minutes

Half life   Half life   Half life

half-life The time taken for half the atoms in a sample of a *radioisotope to *decay. E.g. the half-life of uranium-235 is 700 million years and of radon-222 is 4 days.

halide 1. A *halogen atom that has become an *ion, e.g. fluoride (F^-), chloride (Cl^-), bromide (Br^-), iodide (I^-). 2. A *compound that contains a halogen. Metal halides are usually *ionic, e.g. Na^+Cl^-, and

Harmonic

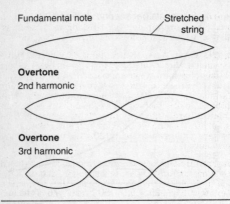

Fundamental note   Stretched string

Overtone
2nd harmonic

Overtone
3rd harmonic

---

non-metal halides are *covalent, e.g. 1,2–dibromoethane

$$Br - \underset{\underset{H}{|}}{\overset{\overset{H}{|}}{C}} - \underset{\underset{H}{|}}{\overset{\overset{H}{|}}{C}} - Br.$$

halogen A *non-metal that is a member of group VII in the *periodic table of the *elements. The halogens exist as molecules and include *fluorine (F_2), *chlorine (Cl_2), *bromine (I_2), and *iodine (I_2). Their reactivity decreases down the group. (see *halide)

haploid number The number of chromosomes in a cell nucleus where there is a single set of unpaired chromosomes. E.g. *gametes (sperms and eggs) are haploid. The normal (*diploid) number is restored during fertilization when sperm and egg fuse together.

hardware The electronic and mechanical parts that make up a a *computer. E.g. *disk drive, *visual display unit, *central processing unit. (see *software)

hard water Water that contains dissolved calcium or magnesium *salts that form scum with *soap. Temporary hard water contains calcium or magnesium *hydrogencarbonate which decomposes during boiling to form solid carbonate limescale. This process makes the water less hard (more soft). Permanent hard water contains calcium or magnesium sulphate. It cannot be softened by boiling. Both types of hard water can be softened by the process of *ion exchange.

harmonic One of the extra *frequencies produced by a musical instrument or oscillator, that are simple multiples of the main (fundamental) frequency. Different harmonics give each instrument its own distinctive sound.   ✍

heart The muscular *organ that pumps blood around the *circulation of an animal. The human heart contains four main chambers (spaces), the left and right *atria and the left and right *ventricles. Contractions of the heart *muscle pump blood through these chambers. The mitral (bicuspid) and tricuspid heart valves control the direction of blood flow. The sound of a heartbeat is caused by these valves closing. Blood from the body returns to the heart through the vena cava. The right atrium and ventricle pump this blood through the pulmonary artery to the *lungs. Oxygenated blood returns from the lungs through the pulmonary vein. The left atrium and ventricle pump the blood through the aorta to *capillaries in the body.   ✍

Heart

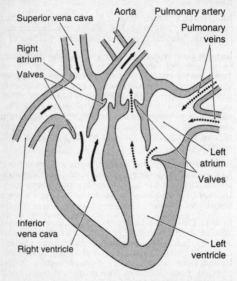

Superior vena cava
Aorta
Pulmonary artery
Pulmonary veins
Right atrium
Valves
Left atrium
Valves
Inferior vena cava
Right ventricle
Left ventricle

heat *Thermal energy contained in an object that has a higher *temperature than its surroundings. Heat can move from a hotter (high-temperature) to a cooler (lower-temperature) part of a solid by *conduction. E.g. a metal rod with one end held in a flame. Heat can move through a *fluid by *convection. E.g. hot air rising from a radiator. Heat can move through empty space by *radiation. All hot objects emit *infra-red radiation (radiant heat).

heat capacity The heat energy needed to raise the temperature of something by 1 °C. The *specific* heat capacity is the heat energy needed to raise the temperature of 1 kg of a substance by 1 °C. The specific heat capacity of water is 4200 J/kg/°C.

heat energy Another name for *thermal energy.

heat pump A machine that absorbs *heat energy from one place and gives it out at another. Heat pumps are used in some advanced heating systems that draw heat energy from the ground or from a river to heat a house. A *refrigerator is a type of heat pump.

heavy water Water that contains *deuterium instead of ordinary hydrogen. It is used as a moderator in some *nuclear reactors.

helium (symbol He) A gaseous *element belonging to group VIII of the *periodic table (the *noble gases). Helium is a monatomic gas, composed of separate atoms. It is extracted from some *natural gas deposits and is used in breathing mixtures for deep-sea divers, in lighter-than-air balloons, and in the manufacture of *semiconductors. Liquid helium cools *superconductors to below 4° K.

heptane see *alkane

herbaceous Describing a plant with soft non-woody tissues.

herbicide A chemical substance that kills plants. Modern herbicides can be designed to kill only certain types of plants. Farmers use herbicides to kill weeds growing among their crops.

herbivore A type of animal that eats only plants. Herbivores have large molar teeth for grinding food and many have no *canine teeth. Their *alimentary canals are very long and contain *bacteria that help to digest *cellulose. Cows, horses, and elephants are herbivores.

heredity The study of the inheritance of *characteristics from one *generation of organisms to the next. (see *gene *allele)

heroin A highly addictive *drug made from sap collected from a type of poppy. It is a narcotic drug that causes drowsiness.

hertz (symbol Hz) The *SI unit that measures the *frequency of a vibration or *wave. E.g. if a string in a guitar vibrates 1000 times each second, then it produces sound-waves with a frequency of 1000 Hz.

heterozygous Describing an organism that has a *dominant and a *recessive allele controlling a *characteristic. A child is heterozygous for the characteristic of brown eyes if it has inherited a brown eye (dominant) allele from one parent and a blue eye (recessive) allele from the other.

hexane see *alkane

HIV Abbreviation for human immunodeficiency virus. (see *AIDS)

hologram A sort of photograph that forms a three-dimensional *image when light is reflected from its surface. The image is formed by *interference between the reflected light waves. Holograms are included on credit cards to make forgery difficult.

homoeostasis The regulation by an organism of the conditions inside it. Homoeostasis keeps the conditions constant. In mammals, *lungs control the amount of carbon dioxide and oxygen in *tissue fluid. The *liver and *pancreas regulate the concentration of glucose in the blood. *Kidneys and sweat glands control the amount of water and salts in the body. Skin and hair help to regulate temperature.

homoiotherm (warm-blooded animal) An animal
that has a steady internal body temperature. Mammals (temperature 36–8 °C) and birds (temperature
38–40 °C) are homoiotherms. Heat made by *respiration in body cells replaces heat lost through the skin.

homozygous Describing an organism that has two
identical *genes controlling a *characteristic. The
genes can both be either *recessive or *dominant.

Hooke's law A law which states that when an elastic object (e.g. a rubber band) is stretched, the
increase in length is proportional to the force
stretching it.

hormone A substance that is made in one part of a
vertebrate animal and has an effect in another part.
Hormones are made by *endocrine glands and circulate in the blood stream. They control and co-ordinate processes such as growth and reproduction.
Each hormone affects a certain part of the organism.

humerus The bone between elbow and shoulder in
humans and many other vertebrates. (see *skeleton)

humidity A measurement that describes the
amount of water vapour in the atmosphere. *Absolute
humidity* is measured in kilograms of water per
cubic metre of air. The maximum humidity of cold
air is less than that of warm air. Rainfall results
when warm humid air cools.

humus A part of *soil, made up from the decayed
remains of plants and animals. Humus provides
many *nutrients for plants and helps soil hold water.

hybrid 1. An organism that has two different alleles
for one *characterstic. One of the alleles is *dominant and the other is *recessive. (see *heterozygous)
2. A hybrid plant or animal comes from parents with
very different characteristics. If the parent organ-

isms are from the same *species, the hybrid off-spring are usually healthier and stronger. Hybrids from parents of different species are usually *sterile (e.g. mules).

hydrated Describing a substance that contains water in its *structure. A crystal of blue hydrated copper sulphate ($CuSO_4.5H_2O$) is a *lattice made up from Cu^{2+} ions, SO_4^{2-} ions, and water molecules. Hydrated substances usually lose water when heated. (see *anhydrous)

hydraulic Describing a *machine that moves mechanical energy from one place to another

Hydraulic

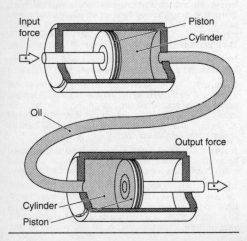

Input force

Piston

Cylinder

Oil

Output force

Cylinder

Piston

through a liquid. Hydraulic machines include vehicle *brakes, earth-moving machines, and hydraulic jacks. They are designed to have large *mechanical advantages to develop very large forces that move heavy *loads. ✎

hydrocarbon A *compound made up from carbon and hydrogen atoms only. Hydrocarbons can be divided into four main groups, *alkanes, *alkenes, *alkynes, and *arenes.

hydrochloric acid A strong *acid with the formula HCl, made by dissolving *hydrogen chloride gas in water. It is an important *reagent in the chemical industry.

hydroelectricity Electrical energy from *generators powered by *turbines that are driven by moving water. The water is usually stored behind a

Hydroelectricity

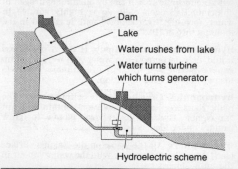

— Dam
— Lake
— Water rushes from lake
— Water turns turbine which turns generator

Hydroelectric scheme

dam built across a river valley. (see *renewable energy) ✍

hydrogen (symbol H) The simplest and lightest of all *elements; atomic number $Z = 1$. Its nucleus contains 1 *proton and it occurs on Earth in *compounds such as water (H_2O) and *methane (CH_4). As well as ordinary hydrogen (sometimes called *protium) there are two other *isotopes of hydrogen, *deuterium and *tritium.

hydrogenation A *chemical reaction where hydrogen joins with a compound. A common example is the addition of hydrogen to *unsaturated vegetable *oils to make solid margarine.

hydrogencarbonate An *ion and an acid *radical with the formula HCO_3^-. Hydrogencarbonates form solid *salts with *alkali metals. The salts are very soluble in water and give carbon dioxide when heated or added to an acid.

hydrogen chloride A fuming acrid colourless gas with the formula HCl, made by burning hydrogen in chlorine. Hydrogen chloride is highly soluble in water, forming *hydrochloric acid. It is used in the manufacture of *PVC.

hydrogen peroxide (formula H_2O_2) A colourless liquid that decomposes to water and oxygen. Manganese dioxide and *catalase are suitable *catalysts to speed up the decomposition.

hydroponics Growing plants without soil. The plant roots are held in a solution containing plant *nutrients. Hydroponics allows plants to grow rapidly under ideal conditions.

hydrosphere All the water on the Earth's surface (see *lithosphere), together with the water vapour in the *atmosphere.

hydroxide A compound that contains the hydroxide ion OH⁻ combined with a metal. E.g. sodium hydroxide NaOH; calcium hydroxide Ca(OH)$_2$. (see *alkali)

hydroxonium ion The *ion (formula H_3O^+) that forms when an *acid dissolves in water. A *proton from the acid joins with a molecule of water, e.g. $HNO_3(l) + H_2O(l) \rightarrow H_3O^+(aq) + NO_3^-(aq)$.

hydroxyl ion The *ion that results when a *base dissolves in water, e.g.

$$NaOH(s) + H_2O \rightarrow Na^+(aq) + OH^-(aq);$$
$$NH_3(g) + H_2O \rightarrow NH_4^+(aq) + OH^-(aq).$$

(see *alkali)

hygroscopic Describing a solid that absorbs water vapour from the air around it. Delicate instruments are packed with cloth bags containing hygroscopic silica gel to decrease the *humidity of the air and to help protect against *corrosion.

hyphae (sing. hypha) The fine threads that make up all the parts of a *fungus. A loose mat of hyphae called the mycelium gives out enzymes that digest food for the fungus. Tightly packed hyphae make up the reproductive organs (e.g. toadstools) that produce *spores.

hypothalamus A part of the brain of a *vertebrate that controls *osmoregulation, hunger and thirst, body temperature, and sleeping.

hypothermia Having an abnormally low body temperature. Hypothermia results (most easily in very young and old people) when body heat is lost to the surroundings faster than it can be made. Body processes slow and unconsciousness or death can result.

hypothesis An explanation for an observation. E.g. observation: 'Objects fall to the ground'; hypothesis:

'There is a force of attraction between the Earth and other masses.' A hypothesis can be tested by *experiments, the results of which can lead to a fuller understanding, a *theory, or even a *law.

IC Abbreviation for *integrated circuit.

igneous rock A type of *rock, formed when *magma solidifies. Igneous rocks are usually very hard and crystalline. Examples include *granite and *basalt.

ileum The part of the small *intestine that is furthest from the *stomach in mammals. The ileum is lined with *villi which help the blood to absorb substances from digested food. It joins on to the large intestine. (see *alimentary canal *caecum)

image The picture formed by a *lens, mirror, or other optical instrument. An image that can be formed on a screen (e.g. by a camera lens on the film) is called a *real image*, because rays of light meet at the image. An image that cannot be formed on a screen (e.g. from a magnifying glass or *concave lens) is called a *virtual image*. Rays of light do not meet at the image.

imago The fully-grown adult form of an insect. Complete *metamorphosis changes a *pupa into an imago. Incomplete metamorphosis changes a *larva into an imago.

immune response see *antigen

immunity The ability of an animal to resist invasion by *antigens that cause a disease. *Natural acquired immunity* results when an animal has suffered from the disease. *Artificial acquired immunity* results when the animal has been *immunized against the disease. *Antibodies present in the blood attack the antigens and stop the disease from developing.

immunize (vaccinate) To inject a vaccine or *serum into an animal to give it *immunity against a disease. A vaccine usually contains *antigens from disease-causing microbes that have been made harmless. The antigens cause *lymphocytes to make *antibodies. A serum contains antibodies which give immediate but shorter-lasting protection. Babies are immunized against measles, mumps, polio, tetanus, and whooping cough.

impermeable, impervious Describing a solid that does not allow a *fluid to pass through. E.g. a plastic sheet is impervious to water.

impure An impure substance contains a small amount of another (usually unwanted) substance or substances mixed with it.

incisor A type of *tooth used for cutting off pieces of food. Incisor teeth are flat and chisel-shaped and are found at the front of the mouth.

inclined plane A slope that can be used as a simple *machine. Pushing a load up an inclined plane requires less effort than lifting the load vertically straight upwards (but the total amount of *work done is the same in both cases).

incus see *ear

indicator A substance whose colour gives information about another substance. Acid–base indicators are dyes which change colour as *pH changes and include litmus (red = acid; blue = alkali), phenolphthalein (colourless = acid; purple = alkali), and universal indicator (red = pH 1; orange = pH 3; yellow = pH 5; green = pH 7; turquois = pH 9; blue = pH 11; purple = pH 13).

induction (electromagnetic induction) Generating an *electric current in a wire by changing the *magnetic field through it. Induction is used in *generators and *transformers. A tape cassette player uses induction to play back when tiny magnets in the tape move past a coil. (see *Fleming's right-hand rule) ✍

inert Describing a substance that does not easily take part in *chemical reactions. E.g. the *noble gases, and metals such as gold that are low in the *reactivity series of metals.

inert gas Another name for *noble gas.

inertia A property of any object that causes it to resist attempts to change its *velocity. The inertia of a spacecraft keeps it travelling at a constant velocity. To accelerate a car requires energy from the engine to work against the car's inertia.

infra red (IR) Invisible *electromagnetic radiation present in sunlight. Any hot object emits radiant heat energy as IR radiation. It is a band of frequencies immediately below the band of visible light in the *electromagnetic spectrum.

infrasound Soundlike *waves with *frequencies too low for human ears to detect (i.e. below 20 Hz). Earthquake waves cause infrasound waves that some animals can detect.

inorganic compound Originally defined as a *compound not resulting from a living process. It is now defined as any compound not containing *carbon atoms held together by *covalent bonds. Examples include water (H_2O), sulphuric acid (H_2SO_4), and common salt (NaCl). Inorganic compounds also include carbon-containing compounds such as *carbonates (e.g. sodium carbonate Na_2CO_3), carbon dioxide (CO_2), and carbon monoxide (CO).

Induction

Moving magnets

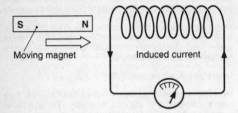

Moving wires

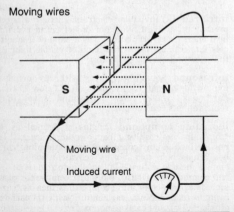

insect A type of *arthropod animal. Insects include flies, beetles, locusts, wasps, and mosquitoes. The body is in three parts: the head, the *thorax, and the *abdomen. Legs and wings are attached to the thorax. All insects have three pairs of legs and most have two pairs of wings. (see *life cycle *metamorphosis)

insecticide A chemical substance that kills insects and other *arthropod animals. Insecticides kill arthropods that damage crops or spread diseases. Modern insecticides can be designed to kill only certain types of arthropods.

insoluble The opposite of *soluble, describing a substance that does not dissolve in a given liquid. E.g. salt is insoluble in *ethanol.

insulation 1. A layer of material (usually plastic) that is a non-*conductor and covers a wire carrying an *electric current. Insulation prevents *short circuits between wires and protects people against electric shock. 2. A layer of material (usually containing trapped air) that is a non-conductor of *heat. Insulation stops heat escaping from hot-water tanks and stops heat entering refrigerators.

insulator A substance through which *heat or an *electric current cannot easily flow. Most *non-metals are good insulators. (see *conductor *insulation)

insulin A *hormone made by *endocrine glands in the *pancreas. It helps to regulate the amount of *glucose in the blood. Insulin causes the *liver to remove glucose from the blood and store it as *glycogen. (see *diabetes)

integrated circuit (IC) An electronic device containing a complicated circuit on the surface of a silicon *chip. The circuit may contain many thousands of *transistors, *diodes, and *resistors. Integrated

Interference

Waves add

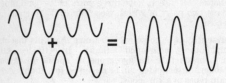

Waves cancel

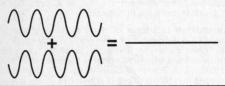

circuits are used in electronic equipment (e.g. *computers, radios, television sets).

interference Combining two *waves together to produce a single wave. Waves can add together or cancel, depending on whether they are in or out of step (phase) with each other.   ✍

internal combustion engine A type of engine that burns its fuel in contact with its moving parts to produce *mechanical energy. Examples include petrol, diesel, and *turbojet engines (but not *steam engines; they burn their fuel in a separate furnace). Four-stroke petrol engines produce power from four movements (strokes) of a *piston: (1) a mixture of air and petrol vapour enters the *combustion chamber;

Internal combustion engine

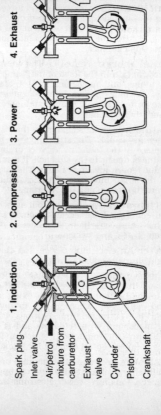

1. Induction   2. Compression   3. Power   4. Exhaust

Spark plug
Inlet valve
Air/petrol mixture from carburettor
Exhaust valve
Cylinder
Piston
Crankshaft

(2) mixture compressed; (3) mixture ignited by an electric spark, burns, expands, and forces the piston down; (4) burned mixture pushed out. The moving piston turns a *crankshaft which drives a *flywheel. *Valves control the flow of gases into and out of the combustion chamber.   ✍

internal energy (symbol U) The energy that substances contain in the form of the *kinetic energy of their atoms and molecules and of the *potential energy in their *bonds. Internal energy decreases during an *exothermic change and increases during an *endothermic change.

intestine A long muscular tube that is part of the *alimentary canal. *Digestion and absorption of food take place in the intestine. In vertebrates, the small intestine connects to the stomach. It is made up from the *duodenum and the *ileum. The large intestine joins the small intestine at the *caecum. It is made up from the *colon and the *rectum.

invertebrate An animal without a backbone (*vertebral column). Invertebrates include sponges, sea-urchins and starfish, *molluscs, *arthropods, *coelenterates, and *worms.

inverter An electronic device that changes direct current into alternating current (*electric current).

iodine (symbol I) A soft grey-black solid *non-metal *element that is a member of group VII of the *periodic table (the *halogens). Iodine exists as molecules I_2 and *sublimes when heated to give a purple vapour. It is extracted from solutions of iodide *salts by displacement with *chlorine and is used in the manufacture of some *antiseptics and medicines. Iodine is an *essential element for humans.

ion An atom or group of atoms that has gained an overall *charge either by taking in or by losing *electrons. *Cations have fewer electrons than *protons

Ion exchange

Hard water
with
Ca^{2+} and
Mg^{2+} ions

— IN →

Ion
exchange
resin

Soft water
with
Na$^+$ ions

— OUT→

and thus have an overall net positive charge.
*Anions have more electrons than protons and thus
have an overall net negative charge. An ion can be a
single atom (e.g. Na$^+$, Cl$^-$) or a group of atoms (e.g.
NH$_4^+$, CO$_3^{2-}$).

ion exchange A process used to remove certain
*ions from a *solution. An ion exchange resin traps
the ions in the solution and exchanges them for
other sorts of ions. Ion exchange is used to make
*hard water soft and to separate dangerous sub-
stances in *nuclear waste (*radioactive waste).

ionic bond (electrovalent bond) The *electrostatic
force of attraction between two neighbouring ions
formed as the result of electron transfer. Metal atoms
transfer electrons to non-metal atoms. Ions form
with the same stable *electronic configurations as
*noble gases. The ions attract each other and form a
*lattice.

ionic compound A *compound that contains *ions
formed from metal and non-metal elements, e.g.
sodium chloride NaCl (Na$^+$Cl$^-$). Ionic compounds
have high melting- and boiling-points and they conduct electricity when molten.

Ionic bond

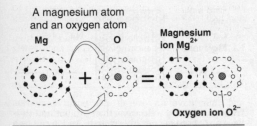

A magnesium atom
and an oxygen atom

Mg   O   Magnesium
ion Mg^{2+}

Oxygen ion O^{2-}

ionic equation A *chemical equation that includes only the *ions involved as *reactants in a *chemical reaction. E.g. chemical equation: $NaOH(aq) + HCl(aq) \rightarrow NaCl(aq) + H_2O(l)$. Ionic equation: $H^+(aq) + OH^-(aq) \rightarrow H_2O(l)$. The spectator ions $Na^+(aq)$ and $Cl^-(aq)$ are not included because they are not involved in any change.

ionization The process of producing *ions. Ionization happens when some molecules dissolve in water, e.g. $HCl \rightarrow H^+ + Cl^-$ and when atoms join to form ionic compounds, e.g. $Na + Cl \rightarrow Na^+ Cl^-$. *Thermal energy, *electromagnetic radiation, and *ionizing radiation can also cause ionization by electron loss: $X \rightarrow X^+ + e^-$. Ionizing radiation can cause ionization by electron gain: $Y + e^- \rightarrow Y^-$.

ionizing radiation *Radiation with sufficiently high energy to cause *ions to form in the substance it is passing through. Ionizing radiation can be a stream of high-energy particles (e.g. *electrons/*beta particles, *alpha particles, *protons) or short-wavelength *electromagnetic radiation (*ultraviolet or *gamma rays, *X-rays). Small amounts cause *mutations in the cells of living organisms. Large amounts cause illness and death.

ionosphere Layers in the atmosphere (between 50 and 1000 km above the Earth's surface) that contain charged particles (*ions). The particles result from the ionization of air molecules by *solar radiation. Radio waves with frequencies between 3 and 30 MHz can be reflected by the ionosphere, allowing long-distance radio communication.

IR Abbreviation for *infra red.

iris see *eye

iron (symbol Fe) A silvery-white *transition metal *element. Impure *pig iron is extracted from iron *ore (see *haematite) in a *blast-furnace. The pig iron is used to make different types of *steel.

isobar A line on a map joining places where the *atmospheric pressure is the same. Isobars show the regions of high and low pressure used to forecast the weather.

isomers Two or more *compounds that have the same molecular *formula but different structural formulae. E.g. there are two isomers of butane C_4H_{10}:

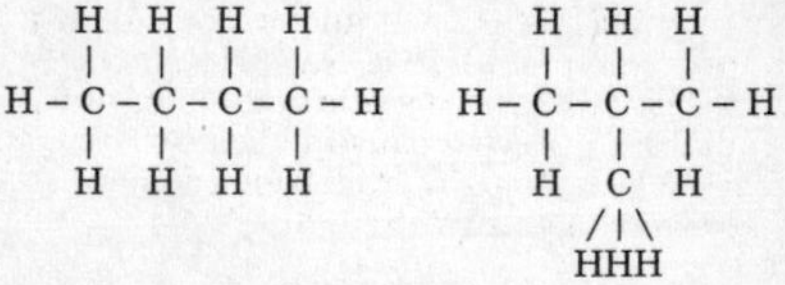

isotopes Two or more forms of the same *element whose atoms have different *mass numbers. E.g. carbon-12 and carbon-14 are isotopes of carbon. They have the same atomic numbers and so the same numbers of electrons and protons but different numbers of neutrons. They also have the same chemical properties.

J

jet engine see *turbojet engine

jet propulsion Moving an object in one direction by using the *reaction force of a high-speed stream of gas or liquid moving in the opposite direction. Swimming octopuses, firework rockets, and *rocket and *turbojet engines use jet propulsion.

joint The place where two or more bones touch. The bones cannot move in fixed joints. The bones in the top of the human skull are held together by fixed joints. There are slightly movable joints between the vertebrae in the *vertebral column (spine). Freely movable joints include the human elbow (hinge joint) and hip (ball-and-socket joint).

joule (symbol J) The *SI unit of *energy and *work. 1 J of work is done when a force of 1 N moves 1 m. 1500 J of heat energy will boil a teaspoonful of cold water. A torch bulb uses 1 J of electrical energy every second. The food eaten by a 15-year-old girl in a day should contain about 10 000 000 J of chemical energy.

K

KE Abbreviation for *kinetic energy.

kelvin (symbol K) The *SI unit of *temperature.

kerosene A liquid containing *hydrocarbons with 10 to 16 carbon atoms per molecule. It is made from *crude oil and is used for heating and as a fuel in *turbojet engines.

kidney The main *organ used by *vertebrates for *excretion and *osmoregulation. Mammals have a pair of kidneys. Each kidney contains about a million nephrons which remove *urea and excess water and *salts from the blood. Groups of nephrons make up the outer cortex and the pyramids in the inner medulla. In the cortex, water and soluble substances pass from the glomerulus into Bowman's capsule. This liquid passes through tubules and the loop of Henle in the medulla where water and useful substances return to the blood. The resulting *urine collects in the pelvis of the kidney and flows through a *ureter and into the *bladder. ✍

kilogram (symbol kg) The *SI unit of mass. E.g. 1 dm of water and a standard packet of sugar each have a mass of 1 kilogram.

kilowatt-hour (symbol kWh) A unit that measures the amount of electrical energy used.

Energy	=	power	×	time
(kWh)		(kilowatts)		(hours)

A 2-kilowatt heater running for 4 hours uses 8 kilowatt-hours of energy. (see *watt)

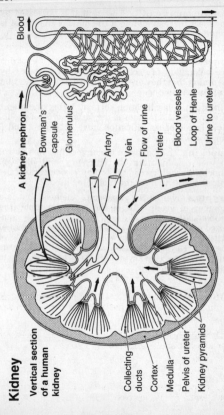

Kidney

Vertical section of a human kidney

Collecting ducts
Cortex
Medulla
Pelvis of ureter
Kidney pyramids

Artery
Vein
Flow of urine
Ureter
Blood vessels
Loop of Henle
Urine to ureter

A kidney nephron

Blood
Bowman's capsule
Glomerulus

kinetic energy

kinetic energy (KE) *Energy that is possessed by an object that is moving. If the mass of the object is *m* and its velocity is *v*, then

$$KE = \tfrac{1}{2}mv^2.$$

kinetic theory (of matter) A *theory that explains the behaviour of *solids, *liquids, and *gases by assuming that they are made up from *particles that are always moving.

kingdom *Classification sorts all living organisms into groups. A kingdom is the largest type of group. Most forms of classification have between 2 and 5 kingdoms.

knocking The explosive burning of fuel in a petrol engine, causing loss of *efficiency and damage. Knocking results when the *compression ratio of the engine is too high for the fuel being used.

krypton (symbol Kr) A gaseous *element that is a member of group VIII of the *periodic table (the *noble gases). Krypton is a monatomic gas, composed of separate atoms. It is extracted by *fractional distillation of *liquid air, which contains 0.0001% of the gas. Krypton is used in powerful *discharge lamps. Radioactive krypton-81 is used in medicine to check lung function.

L

lactation The production of milk by *mammary glands, after the birth of young. Lactation continues for as long as the milk is taken away.

lacteal see *villus

lactic acid A substance made in *muscle tissue during *anaerobic respiration when there is a shortage of oxygen. It can build up in a muscle to give cramp-like pains. It is either changed into *glucose or broken down by the *liver.

Lamarckism A theory of evolution put forward by Lamarck in 1809. It suggests that changes that happen to an organism during its lifetime are inherited by its offspring. E.g. giraffe ancestors stretched their necks to reach the tops of trees, so each new generation was born with longer necks. *Darwinism has replaced Lamarckism.

larva One feeding stage in the *life cycle of some insects and other *invertebrates. The larva looks completely different from the adult. A caterpillar is the larva of a butterfly. (see *metamorphosis)

larynx The top part of the trachea (*lung) in most *vertebrate animals. The larynx in mammals, reptiles, and amphibians contains the vocal cords that make sound when air passes over them.

laser An electrical device that gives out a beam of *coherent light that is very narrow and intense. Lasers are used to carry out delicate operations on eyes and nerves. They are also used in compact disc players and in automatic supermarket check-outs (to read *bar codes).

latent heat The heat energy given out or taken in as the *state of a substance changes at a constant temperature. E.g. the latent heat of fusion of water is 336 joules per gram (J/g). This is the energy given out when 1 g of water at 0 °C freezes into ice at 0 °C. The same amount of energy is taken in when 1 g of ice at 0 °C melts to water at 0 °C. The latent heat of vaporization of water is 2260 J/g. This is the energy taken in when 1 g of water at 100 °C changes to steam at 100 °C. The same amount of energy is given out when 1 g of steam at 100 °C condenses to water at 100 °C.

lattice In *crystals, the regular arrangement of *atoms, *ions, or *molecules in a repeating pattern. (see *giant structure)

lava see *magma

law A rule that is true everywhere and can predict the outcome of observations before they are made. E.g. *Ohm's law predicts that doubling the voltage across any conductor will double the electric current flowing through it (provided the temperature is kept constant).

LDR Abbreviation for *light-dependent resistor.

leaching The washing away of plant nutrients, usually by rain-water draining quickly through the *soil. (see *soil depletion)

lead (symbol Pb) A soft dense grey *metal *element that is a member of group IV in the *periodic table. Lead is extracted by *smelting the ore (*galena) in air and then with carbon. It is used in car batteries and alloys (e.g. *solder, pewter) and in sheets that help to waterproof roofs. Lead and its *compounds are poisonous. (see *tetraethyl lead)

lead–acid cell A type of *secondary cell made from two lead plates immersed in sulphuric acid. *Batter-

Leaf

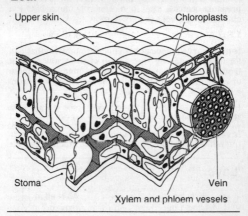

ies made up from lead–acid cells are used in cars and
lorries.

leaf A structure growing from a *stem on a plant.
Most leaves are flat and broad so that they have a
large surface area. Green leaves contain *chlorophyll used in *photosynthesis. The skin (especially
of the lower sides) contains holes called *stomata. ✍

Le Chatelier's principle A rule that states: 'If the
conditions of a *chemical equilibrium are changed,
the equilibrium will alter so as to counteract the
change.' E.g. nitrogen dioxide NO_2 and dinitrogen
tetraoxide N_2O_4 are in equilibrium: $N_2O_4(g) \Leftrightarrow
2NO_2(g)$. Adding N_2O_4 causes the formation of more
NO_2, which uses up N_2O_4. Raising the temperature

causes more N_2O_4 to *decompose, an *endothermic change that lowers the temperature. Increasing the pressure causes the formation of more N_2O_4, a change that lowers the pressure.

LED Abbreviation for *light-emitting diode.

lens A curved piece of transparent material that can form an *image by *refraction of the *rays of light from an object. Lenses are usually made from glass or clear plastic and include *convex lenses and *concave lenses. (see also *eye)

leucocyte A white *blood cell. (see *phagocyte *lymphocyte)

lever A simple *machine made of a bar and a turning-point (*fulcrum, *pivot). *Effort pushing on one

Lever

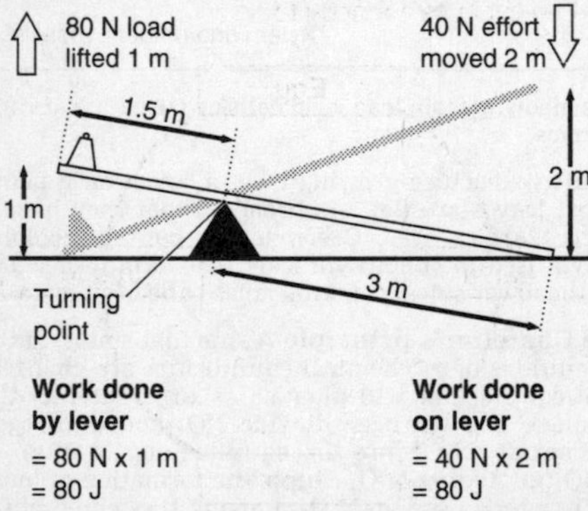

Work done by lever	Work done on lever
= 80 N x 1 m	= 40 N x 2 m
= 80 J	= 80 J

end of the bar moves a *load attached to another part of the bar. (see *mechanical advantage *velocity ratio)   ✍

LH Abbreviation for *luteinizing hormone.

lichen An organism made up from an *alga and a *fungus living closely together in *symbiosis. The alga uses *photosynthesis to make food for the fungus. The fungus protects the alga.

life cycle The main stages that a plant or animal passes through during its life. The human life cycle includes the stages: *egg–*embryo–*foetus–birth–baby–child–adolescent–adult–old age–death. The life cycle of *insects and *amphibians involves *meta

Life cycle

Life cycle of a mosquito

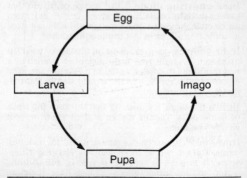

morphosis, when body shape changes completely. (see *larva *pupa *imago *nymph) ✍

ligament A tough piece of *tissue that holds two bones together at a joint. Elbow and knee joints contain ligaments.

light *Electromagnetic radiation that can be detected by the eyes of an animal. White light is a mixture of different colours and can be split into a *spectrum of red, orange, yellow, green, blue, and violet light. Each colour has its own *wavelength in the *electromagnetic spectrum.

light bulb A device that changes electrical *energy into light and heat energy. An *electric current flows through a thin coil of *tungsten wire (called the filament) that glows white-hot. The glass bulb is filled with inert argon and nitrogen gas that stops the filament burning away.

light-dependent resistor (LDR) A *resistor whose resistance falls when light shines on it.

light-emitting diode (LED) A type of *diode that gives out light (usually red, green, or yellow) when an electric current flows through it. LEDs are used as indicator lamps in electronic equipment.

light-year A measurement of distance used in astronomy. A light-year is the distance travelled by a beam of light in 1 year, equal to 9.465×10^{12} km. The distance from the Sun to the closest star is approximately 4.3 light years.

lignin In plants, a substance that thickens the walls of some cells. Lignin makes a plant stem or root woody and hard.

lime (quicklime) The common name for calcium oxide, CaO, a white solid made by roasting *limestone. It is used to *neutralize acid in soil. Adding water to quicklime makes slaked lime (calcium

hydroxide), a white solid used in glassmaking and
*mortar, as well as to neutralize acid soil.

limescale see *hard water

limestone A *mineral that contains mostly calcium
carbonate $CaCO_3$. It is used to remove impurities
during the extraction of *iron in a *blast-furnace,
and in the manufacture of *cement and quicklime
(*lime).

lines of force Lines drawn to show the shape and
direction of the *magnetic field around a *magnet.
By convention, lines of force run from the north pole
to the south pole of a magnet. They can be plotted by
using a small compass. The lines of force are closest
together where the field is most intense (strongest).

lipid In living organisms, any member of a large
group of substances that do not dissolve in water.
Examples include *fats and *oils used as food stores
and *waxes used for waterproofing. Lipids are also

Lines of force

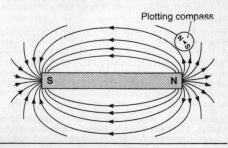

Plotting compass

found in *cell membranes and in some *vitamins and *hormones.

liquid A *state of matter where a substance is a runny *fluid. A liquid sinks to cover the bottom of its container and has a flat surface. The *particles in a liquid are closely packed but are able to move freely past each other.

liquid air A liquid (boiling-point about −200°C) made by cooling and compressing air. *Fractional distillation separates *oxygen, *nitrogen, and *noble gases from liquid air.

lithium (symbol Li) A soft silvery *metal *element that is a member of group I in the *periodic table (the *alkali metals). Lithium is extracted by *electrolysis of molten lithium chloride. It is a very reactive element and is used in special types of electric *battery.

lithosphere The Earth's solid crust. It is 8–70 km thick and covers the inner molten parts. The word 'lithosphere' sometimes includes the *mantle, and occasionally the core as well. (see *hydrosphere)

litmus see *indicator

litre (symbol l) A unit of volume. 1 l is the *volume of a cube with sides 10 cm. A large bottle of lemonade holds 2 l.

liver A large organ in the abdomen of vertebrates. The human liver stores vitamins A, B, and D as well as iron needed to make red blood cells. The liver produces heat that helps to warm the body. It also stores glucose as *glycogen. It changes waste from unwanted *amino acids into *urea and breaks down poisons (including alcohol and drugs). It makes *bile and substances that help blood to clot in wounds.

liverwort A type of plant that grows in damp places. Liverworts do not have flowers or proper roots or leaves. They do not have stems, and look like fleshy leaves growing flat on the ground. Liverworts use *spores to reproduce.

live wire The wire (coloured brown or red) in a mains electricity lead. Electrons in the live wire carry electrical energy from the power-station to an electrical appliance. The voltage of the live wire is 240 V in the UK. The *neutral wire conducts the electrons back to the power-station. The voltage of the neutral wire is usually zero.

load The object that a force works against or which a *machine does work on. (e.g. see *lever)

loam The most *fertile type of *soil, made up mostly from sand, clay, and *humus. Loam forms into tiny clumps called soil crumbs which have air spaces between them. The crumbs soak up water and help to stop *nutrients being washed away.

logic gate A type of electronic switch. A gate has one or more inputs and a single output. An input is

Logic gate

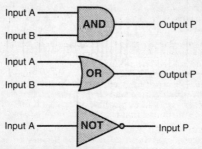

ON when current flows into it. An output is ON when current flows from it. An AND gate has two inputs; the output is ON when both inputs are ON. An OR gate has two inputs; the output is ON when either one input or the other is ON. A NOT gate has one input; the output is ON when the input is not ON (OFF) and vice versa. A *computer uses thousands of logic gates to process *data expressed as pulses of current (*bits). (see *truth table *bistable) ✍

long sight A person with long sight cannot see near objects clearly. The *eye focuses them behind the retina. Long sight is overcome by wearing spectacles with *convex lenses.

loop of Henle see *kidney

Loudspeaker

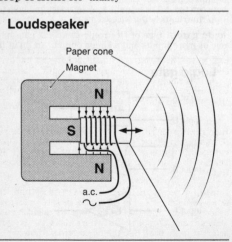

Paper cone

Magnet

N

S

N

a.c.

loudspeaker A device that changes an electric signal into sound-waves. The signal is a rapidly changing *electric current. It flows through a coil of wire in the *field of a powerful *magnet. As the current flows back and forth, the coil is repelled and attracted by the magnetic field. The coil vibrates a paper or plastic cone, making it give out sound-waves. ✍

luminous Describing something that gives out light. E.g. the Sun, a light bulb, a glow-worm.

lung The *organ used for *breathing air. In mammals, two lungs are contained in the *thorax, one each side of the heart. Each lung contains groups of air-filled sacs called alveoli which are surrounded by *capillaries containing blood. Oxygen diffuses from the alveoli into red *blood cells in the capillaries. Carbon dioxide diffuses from the blood *plasma into the alveoli. Breathing causes air to enter and leave the alveoli through bronchioles which connect to the bronchus. The bronchus of each lung joins the trachea which leads to the nose and mouth. The trachea, the bronchi, and the bronchioles are reinforced with rings of *cartilage. They are also lined with *mucus glands and *cilia which help to remove dust. ✍

luteinizing hormone (LH) A *hormone made in the *pituitary gland of a mammal. In males, it helps with the production of *testosterone. In females, it helps with *ovulation and the production of *progesterone.

lymph A liquid that comes from *tissue fluid. It flows through tubes called lymph vessels, carrying dead cells and microbes away from *tissues. Lymph glands make *lymphocytes which clean the lymph before it flows back into the blood *circulation.

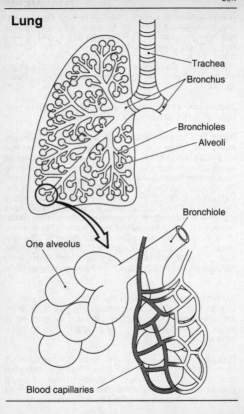

Lung

Trachea

Bronchus

Bronchioles

Alveoli

Bronchiole

One alveolus

Blood capillaries

lymphocyte A type of white *blood cell which makes *antibodies. Lymphocytes help to protect against infection. They are made in parts of the *lymph system called nodes.

M

machine A *device that makes doing work against a force easier. The simplest machines are the *lever, *wedge, *inclined plane, *screw, wheel and *axle, and *pulley. More complicated machines are usually made up from combinations of these. A force (called the effort) applied to one part of a machine results in a force moving a *load at another place. Some machines make the force on the load larger (but the distance it moves less). Some machines make the movement larger (but the force less). (see *velocity ratio *mechanical advantage)

macromolecule A very large molecule, usually an *organic *polymer. *Starch, *haemoglobin, and *polythene are made up from macromolecules.

magma Molten *rock that comes from the Earth's *crust or *mantle. Lava from volcanoes is magma that has forced its way to the surface of the Earth. *Igneous rocks form when magma slowly cools and solidifies.

Magnalium The trade name of a strong and light *alloy that contains *aluminium and *magnesium. It is harder than pure aluminium and is used to make aircraft bodies.

magnesium (symbol Mg) A silvery *metal *element that is a member of group II in the *periodic table (the *alkaline earth metals). Magnesium is extracted by *electrolysis of molten magnesium chloride. It is a reactive element and burns fiercely with a brilliant white flame. It is used to make light alloys (usually with *aluminium).

Magnetic field

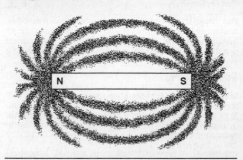

magnet An object that will attract iron and is able to repel other magnets. A magnet has two *magnetic poles. (see *permanent magnet *electromagnet *solenoid)

magnetic field The space around a *magnet where magnetic forces can be detected. The magnetic field can be shown by sprinkling iron filings on to a sheet of paper covering a magnet. (see *lines of force) ✍

magnetic pole A place where the *lines of force of a *magnetic field enter or leave a magnet. A magnet has two poles, a north pole and a south pole. Between separate magnets, like poles (S–S or N–N) repel each other and opposite poles (N–S) attract. The north pole of a compass needle points towards the North Pole—the Earth's magnetic south pole.

magnetron An electronic device that generates *microwaves.

malachite An *ore used in the production of *copper. It contains copper carbonate-hydroxide $CuCO_3.Cu(OH)_2$.

malleability The ability of a metal to be pressed or hammered into a new shape without cracking or breaking. Drink cans are made from aluminium because it is sufficiently malleable not to tear during pressing.

malleus see *ear

malnutrition A lack of proper nourishment suffered by a person who does not have a *balanced diet. It can result from a lack of protein, carbohydrate, fat, or vitamins. Some forms of malnutrition can lead to *deficiency diseases.

mammal A type of *vertebrate animal. Mammals include humans, rats, and walruses. They are warm-blooded (*homoiotherms), have hair, and feed their young on milk. All mammals give birth to live young except duck-billed platypuses and spiny ant-eaters, which lay soft-shelled eggs.

mammary gland In female mammals, a *gland that produces milk. Apes and humans have mammary glands on the *thorax (chest). Other mammals have them on the *abdomen. (see *lactation)

mandible Grasshoppers, crabs, and many other *insects, *crustacea, and *myriapods use a pair of horny mandibles to bite and crush food. The word 'mandible' is also used for the lower jaw of a *vertebrate or either part of a bird's beak.

manometer A device used to measure gas pressure. The gas pushes a liquid around a U-shaped tube until the downward force (from the weight difference between the two liquid levels) equals the upward force (from the gas pressure). The difference between

Manometer

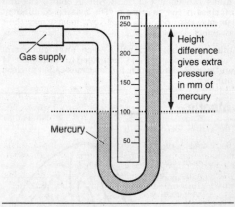

Gas supply

mm
250

Height
difference
gives extra
pressure
in mm of
mercury

200

150

Mercury

100

50

the levels shows how high the pressure of the gas is
compared to *atmospheric pressure.   ✍

mantle The region inside the Earth between the
*crust and the liquid, possibly metallic, core. The
mantle is molten rock, moving constantly in *convection currents.   ✍

marble A type of *rock that contains calcium carbonate and some magnesium carbonate. Pure marble
is white and is easy to cut, carve, and polish.

marrow A soft *tissue that fills the spaces inside
*bones. Yellow marrow consists mostly of cells con-
taining fat. It helps to make some types of white
*blood cells. Long bones, such as human leg bones,

contain red marrow in their ends. Red marrow produces red blood cells.

mass The amount of *matter in an object. The unit of mass is the *kilogram. The masses of different objects can be compared by measuring their *weights.

mass number (symbol *A*) The total number of *nucleons (*protons and *neutrons) in the nucleus of an *atom. Mass number is sometimes called nucleon number.

Mantle

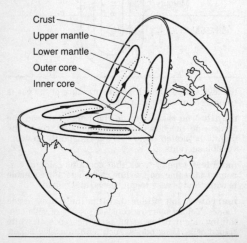

Crust

Upper mantle

Lower mantle

Outer core

Inner core

material Another word for *matter or substance.

matter Anything that has *mass and *volume. All objects are made up from matter. *Solid, *liquid, *gas, and *plasma are the four states (forms) of matter.

measuring cylinder A transparent container used to measure approximately the volume of a sample of liquid. Measuring cylinders are made in sizes that range from 10 cm³ up to 2000 cm³. ✍

mechanical advantage For any *machine (e.g. see *lever), the force moving the *load compared to the *effort used.

$$\text{mechanical advantage} = \frac{\text{load}}{\text{effort}}$$

Measuring cylinder

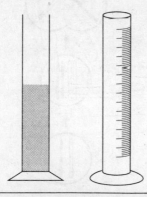

4 gametes

Meiosis

mechanical energy *Energy produced or used by a *machine, where a *force causes part of the machine to move.

medulla 1. A part of the *brain in *vertebrates. The human medulla is at the rear of the brain. It automatically controls heart rate, breathing, and *blood pressure. 2. The central part of an *organ. (see *kidney)

meiosis How *chromosomes replicate and separate in the cells of plant and animal sex organs. It forms sex cells (*gametes) such as *sperms, *ova, and *pollen. During meiosis, a copy is made of each chromosome pair. All the chromosomes then separate to form four sex cells, each with half the number of chromosomes of the original cell. (see *mitosis *haploid number *diploid number)   ✎

melt To change a solid into a liquid by heating.

melting-point (m.p.) The fixed temperature at which a *pure solid melts and changes into a liquid. Melting-point is a *physical property used to identify a solid. E.g. ice, m.p. = 0 °C; iron, m.p. = 1536 °C. Impurities usually lower the m.p. Solid mixtures melt away over a range of temperature.

membrane A thin sheet of *tissue that covers or supports part of a plant or animal. A cell membrane surrounds all living *cells. The tympanum (eardrum) is a membrane in the *ear.

Mendelism A theory of *heredity put forward by Mendel in 1866. The modern version suggests how *characteristics are controlled by *dominant and *recessive alleles.

meniscus The curving of the surface of a liquid where it meets the walls of its container. A meniscus results from the forces of attraction between the liquid and the walls of the container.   ✎

Meniscus

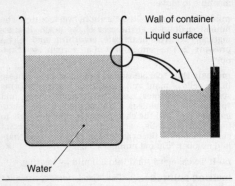

Water

Wall of container

Liquid surface

menstrual cycle A regular monthly change in the reproductive organs of female humans, monkeys, and apes. The lining of the uterus steadily thickens while an ovary releases an *ovum (egg cell). If it does not become fertilized, the ovum dies. The uterus lining then breaks down and flows out through the vagina over a period of 2 to 7 days. Human females call this stage their 'period'. The menstrual cycle is controlled by *hormones made by the ovaries.

mercury (symbol Hg) A dense silvery liquid *transition metal *element. Mercury is made by heating the *ore in air. It is used in thermometers and in *alloys (called amalgams) for dental fillings. Its *compounds are poisonous.

metabolism All the *chemical reactions that happen in a living organism. Metabolism includes breaking down food during *digestion, releasing energy during *respiration, and the *excretion of waste substances.

metacarpal In the human hand, one of the bones that join the wrist bones (carpals) to the finger bones (phalanges). (see *skeleton)

metal A substance whose typical *physical properties are *ductility, *malleability, and good conduction of heat and electricity. A freshly cut metal surface is usually shiny. Many metals are *elements included on the left-hand side and in the centre of the *periodic table. Their typical *chemical properties are the formation of basic *oxides and *cations. Many metals mix to form *alloys (see *alkali metal *alkaline earth metal *transition metal)

metallic bond The *electrostatic force of attraction between two neighbouring *metal nuclei and delocalized *valency electrons. The valency electrons

Metallic bond

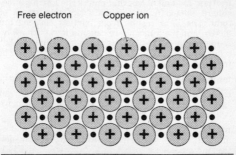

Free electron   Copper ion

move freely and are shared by all the atoms in the
*lattice. ✐

metalloid (semi-metal) An *element whose properties are intermediate between those of *metals and
*non-metals. *Silicon and germanium are metalloids
which are used to make *semiconductors.

metamorphic rock A type of *rock, formed when
*sedimentary rock is heated and pressurized in the
Earth's *crust. Metamorphic rocks are usually hard
and crystalline. An example is *marble, formed from
*limestone.

metamorphosis A complete change of shape of an
animal. Metamorphosis steadily changes tadpoles
into frogs. Flies and mosquitoes undergo complete
metamorphosis from *larva to *pupa and pupa to
adult *imago. Locusts and cockroaches undergo
incomplete metamorphosis. The larva becomes more
like the adult each time it moults its skin.

metatarsal In the human foot, one of the bones that
join the ankle bones (tarsals) to the toe bones (phalanges). (see *skeleton)

meteor A small object that enters the Earth's atmosphere, appearing in the night sky as a streak of
light. The light results from friction between the
meteor and the atmosphere causing it to burn away.
A meteorite is the solid remains of a meteor that
reaches the Earth's surface. Meteorites are made of
iron and *silica rock.

meteorite see *meteor

methane A gas that is a *hydrocarbon and an
*alkane. Its formula is CH_4. Methane is the main
component of *natural gas.

methanoic acid The simplest *carboxylic acid (formula HCOOH), occurring naturally in stinging-net-
tles and ants.

methanol see *alcohol

metre (symbol m) The *SI unit of length. E.g. 50 1p coins laid side by side will give a row 1 metre long.

mica A *mineral that is found in flat transparent sheets. It is an electrical insulator but a good conductor of heat. Mica is used in some electrical heaters and *capacitors.

microbe see *micro-organism

microchip Another name for *chip.

micro-organism (microbe) Any organism smaller than about 0.1 mm. Micro-organisms can only be seen with the help of a microscope. They include *bacteria, *viruses, *protozoa, and some *fungi and *algae.

microphone A device that changes sound waves into a changing *electric current. The *frequency of the current matches the frequency of the sound waves.

microprocessor A small computer containing only a few thousand *logic gates. Appliances such as washing machines and hand-held calculators can be controlled by a microprocessor in a single *integrated circuit.

micropyle A small hole in the outer covering (testa) of a *seed. Water enters through the micropyle and starts the *germination of the seed.

microwaves *Radio waves with a *frequency between 3×10^{11} Hz (300 000 MHz) and 10^{10} Hz (10 000 MHz). Microwaves carry communications around the world via *artificial satellites and microwave relay towers. Microwave ovens use microwave energy to heat food.

milligram (symbol mg) A unit of *mass, equal to one thousandth of a gram.

millilitre (symbol ml) A unit of volume. 1 ml is equal to 1 cubic centimetre (1 cm³), the volume of a cube with sides 1 cm. There are 1000 millilitres in 1 *litre.

mineral A substance found in the ground that can be identified by its *chemical and *physical properties, e.g. *marble, *limestone. (see *rock *ore)

mirror A surface that reflects the light falling on it. A flat (plane) mirror gives an upright unmagnified virtual *image. Convex mirrors bulge outwards. They give a wide angle of view and are used for driving and for security mirrors. Concave mirrors curve inwards. They give a magnified view and are used as shaving and make-up mirrors.

mitochondrion (pl. mitochondria) A type of *organelle found in a living *cell. It uses *respiration to produce energy from substances (usually glucose) taken in from the *cytoplasm.

mitosis How *chromosomes replicate and separate when living cells divide to make copies of themselves. During mitosis, a copy is made of each chromosome pair. The two sets of chromosomes then separate to form two cells, each with the same number of chromosomes as the original cell. (see *meiosis *diploid number *haploid number)

mixture A substance that is not a pure single substance. A mixture is made up from two or more separate substances. The composition may vary but the mixture has no new properties. *Solutions, *suspensions, and *emulsions are mixtures. The substances in a mixture can be separated by methods such as *evaporation, *crystallization, and *filtering that do not involve chemical changes.

moderator see *nuclear reactor

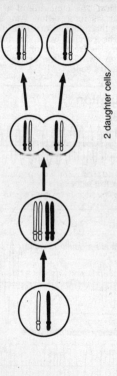

2 daughter cells.

Mitosis

modulation The changing of a *carrier wave so that it carries information. Amplitude modulation (AM) uses the information signal to alter the amplitude of the carrier wave. Frequency modulation (FM) alters the frequency of the carrier wave. Pulse modulation switches the carrier wave on and off according to a code.

Modulation

Carrier

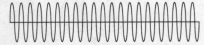

Sine-wave signal

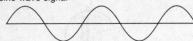

Amplitude-modulated wave

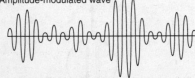

Frequency-modulated wave

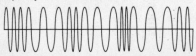

molar 1. A molar solution (symbol 1 M) has a concentration (*molarity) equal to 1 mole of *solute dissolved in 1 dm³ of solution. A solution of half this concentration is 0.5 M. 2. A type of *tooth used for grinding food. Molars are wide and ridged and are found at the back of the jaw. They usually have two or three roots.

molar gas volume At room temperature and pressure, the volume of 1 mole of a gas, equal to about 24 dm³ (24000 cm³). The molar gas volume at *standard temperature and pressure (stp) is 22.4 dm³.

molarity The concentration of a solution calculated as moles of *solute in 1 dm³ of solution. E.g. if a solution of salt contains 58.5 g (1 mole) of sodium chloride (NaCl) in 0.5 dm³ of solution, then the molarity of the solution is 2 moles per dm³.

mole An amount of a substance, equal to its *formula mass expressed in grams. E.g. 1 mole of water (H_2O) is $(1+1+16)g = 18 g$. 1 mole of a substance contains 6.02×10^{23} (*Avogadro's constant N_A) particles of the substance, e.g. 1 mole of water is 18 g and contains N_A H_2O molecules; 1 mole of salt (NaCl) is 58.5 g and contains N_A Na^+ ions and N_A Cl^- ions; 1 mole of iron (Fe) is 56 g and contains N_A Fe atoms.

molecule A small *particle with full electron *shells that is made up from two or more atoms held together by *covalent bonds. A molecule of water is H–O–H, H_2O. A molecule of carbon dioxide is O=C=O, CO_2.

mollusc A type of *invertebrate animal with a soft body and one or two shells. Shellfish and snails are molluscs with external shells. Octopuses, squid, and slugs have internal shells.

moment The turning effect of a force.

Moment = force×distance from the turning-point.

Moments can be clockwise or anticlockwise, depending on which direction they turn. (see *axis *fulcrum)

momentum For a moving object, a quantity calculated by multiplying the object's *velocity by its *mass. E.g. a 5 kg mass moving at 4 metres per second (m/s) has a momentum of 20 kg m/s.

monatomic Describing a substance made up of single separate *atoms of an *element. E.g. *helium is a monatomic gas.

monera A group of very simple *unicellular organisms without a proper nucleus in their cells. It includes *bacteria and blue-green *algae. All monera are members of the monera *kingdom.

monohybrid cross Generating offspring from parents that have one *dominant and one *recessive allele for the *gene controlling a particular characteristic. The ratio of dominant:recessive *phenotype in the offspring is 3:1, i.e. there is a 1 in 4 chance of the offspring not showing the characteristic.

monomer A molecule that can take part in a *polymerization reaction and become part of a *polymer. E.g. *ethene is the monomer used to make *polythene.

moon A natural *satellite that moves around a planet. E.g. Earth has one moon, Mars two, and Jupiter sixteen.

mortar A mixture of *cement, sand, and water that sets and is used to hold together the bricks and blocks in walls and buildings.

moss A type of *plant that usually grows in damp places. Mosses have tiny leaves at the end of thin stems but do not have flowers or proper roots. They use *spores to reproduce.

motor A device that makes *mechanical energy from electrical energy or from the chemical energy in a fuel. (see *electric motor *engine)

mould A simple type of *fungus. Moulds consist of a woolly mass of *hyphae. They are *saprophytes and grow on dead plants and animals.

m.p. Abbreviation for *melting-point.

mucus A slimy substance that protects and lubricates the mucous membranes in an organism. Mucous membranes line organs that join to the outside of the organism, e.g. *intestine, *trachea, *bronchus, *vagina.

muscle A *tissue that can contract, made up from cells called muscle fibres. Pairs of voluntary muscles are attached by *tendons to *joints. They are controlled by the conscious part of the *brain. Layers of involuntary muscles surround blood vessels, gut, uterus, and bladder. They are automatically controlled by the unconscious part of the brain. The hearts of vertebrates contain cardiac muscle. It is a type of involuntary muscle that contracts automatically, without control from the brain. (see *antagonistic pair)

mutagen A chemical substance, virus, or type of *radiation that causes *mutations in living cells. Mutagens change the *DNA in *genes or cause damage to *chromosomes.

mutation A sudden change in a *chromosome or *gene. Mutation alters the way an *embryo develops. Mutations in older organisms can cause the growth of tumours and *cancer. Mutation can happen naturally or be caused by *X-rays, *nuclear radiation, and some chemicals. (see *evolution)

mycelium see *hyphae

myriapod A type of *arthropod animal. Myriapods have bodies made up from segments. Each segment has one or two pairs of legs. Centipedes and millepedes are myriapods.

nasal cavity see *nose

national grid A network of overhead and underground cables used to distribute electrical power throughout the UK. All *power-stations and all consumers are joined to the national grid. If one power-station breaks down, the supply to consumers is not interrupted.

native An *element that occurs free and not as a *compound in its *ore. Native metals are unreactive and include gold, platinum, and silver. *Sulphur is a non-metal that occurs native.

natural gas A gaseous *fossil fuel consisting mostly of *methane. It is used as a *fuel for heating and as a raw material for making other chemical substances. (see *biogas *steam reforming)

nebula A large thin cloud of dust and gas that can exist in space. Gravity may pull a nebula together to contract and to form a star.

nectary A part of some *flowers, producing a sugary liquid called nectar. The nectar attracts insects that pollinate the flower. Flowers with nectaries usually have bright petals.

nematode A type of worm that has a smooth body without segments. Microscopic nematodes help to break down and recycle dead plants and animals. Some nematodes are *parasites that cause diseases.

neon (symbol Ne) A gaseous *element belonging to group VIII of the *periodic table (the *noble gases). Neon is a monatomic gas, composed of separate atoms. It is extracted by *fractional distillation of *liquid air, which contains 0.002% of the gas. *Discharge lamps containing neon glow red and are used for advertising signs and indicator lamps.

nephron see *kidney

nerve In animals, a bundle of nerve fibres. Each fibre is the axon of a nerve cell (*neurone). The nerve fibres are separated from each other by a fatty coating. Nerves can contain fibres from sensory neurones or from motor neurones or a mixture of the two.

nervous system In animals, the cells and tissues that carry information. It controls all the different parts of the body so that they work together properly. The nervous system in vertebrates is made up from the *brain, the *spinal cord, and *nerves.

neurone A *nerve cell. The main part is the nerve fibre, which is made up from a long threadlike axon covered in an insulating coat. A nerve fibre carries impulses in one direction only. Sensory neurones carry information from *sense organs to the *central nervous system. Motor neurones carry impulses from the central nervous system to *muscles and *glands. ✍

neutral 1. Describing a liquid or solution that does not affect litmus *indicator and is neither an acid nor an alkali. Neutral solutions have a pH of 7 because the concentrations of H⁺ and OH⁻ ions are equal. Pure water, sugar solution, and salt (sodium chloride) solution are neutral. 2. Describing an object that is not charged (see *charge *neutron). An atom is neutral when it has the same numbers of *electrons and *protons.

neutralize To react an *acid with a *base to form a solution that is *neutral.

neutral wire The wire (coloured blue or black) in a mains electricity lead. (see *live wire *earth wire)

Neurone

Motor neurone

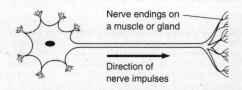

Nerve endings on a muscle or gland

Direction of nerve impulses

Sensory neurone

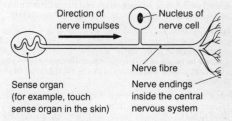

Direction of nerve impulses

Nucleus of nerve cell

Nerve fibre

Sense organ (for example, touch sense organ in the skin)

Nerve endings inside the central nervous system

---

neutrino A particle that has no charge and that travels at the speed of light. Neutrinos are thought to have zero mass when stationary.

neutron A particle contained in the *nucleus of all *atoms except hydrogen. A neutron has no *charge. It has approximately the same mass as a *proton.

neutron star A star that has run out of fuel and collapsed until electrons are forced into atomic nuclei where they combine with *protons and form *neutrons.

If our Sun became a neutron star, its diameter would
shrink from 1 400 000 km to about 8 km and its density
would rise to about 10^{18} kg/m^3.

newton (symbol N) The *SI unit of *force. 1 N gives a
mass of 1 kilogram an acceleration of 1 metre/second/
second (1 ms^{-2}).

Newton's laws of motion Three laws that describe
the effects of forces on objects. 1st law: An object stays
still or continues to move steadily until a force pushes
or pulls it. 2nd law: The rate of change of momentum of
a moving object is proportional to the force acting on it.
Force = mass×acceleration. 3rd law: If object 1 exerts a
force on object 2, then object 2 exerts an equal and opposite force (the *reaction) on object 1.

nicad cell A type of *secondary cell made from a nickel
and a cadmium *electrode in an alkaline *electrolyte.

niche (ecological niche) The relationship that an organism has to its *environment, and the conditions
under which it lives. Each *species of organism can be
defined in terms of its niche. Two different species cannot occupy identical niches at the same time and in the
same place.

nickel (symbol Ni) A hard silvery *transition metal
*element. It is extracted from its *ore by *smelting. It is
used to make stainless *steel and other *alloys. It is also
a *catalyst for *hydrogenation reactions.

nicotine An addictive *drug found in tobacco. It is
used as an *insecticide.

nitrate An *ion and an acid *radical with the formula
NO_3^-. Nitrates form salts, e.g. potassium nitrate KNO_3.
All nitrate salts are soluble in water and decompose
when heated to give oxygen and/or *nitrogen dioxide
gas.

nitric acid A strong *acid with the formula HNO_3,
made by dissolving *nitrogen dioxide gas in water. It is

used in the manufacture of *fertilizers, explosives, medicines, and dyes.

nitrite An *ion and an acid *radical with the formula NO_2^-. Nitrites form salts, mostly with *alkali metals, e.g. potassium nitrite KNO_2. All nitrites are soluble in water and decompose when heated.

nitrogen (symbol N) A colourless gaseous *element that is a member of group V of the *periodic table. It makes up 78% of *air and is very unreactive. Nitrogen exists as molecules N_2, each containing two atoms held together by a strong triple *covalent bond. It is extracted by the *fractional distillation of *liquid air and is used in the manufacture of *ammonia by the *Haber process.

nitrogen cycle The movement of nitrogen between the *atmosphere, substances in the ground, and living organisms. *Bacteria and lightning join nitrogen with oxygen to form *nitrates in the soil. Plants use nitrates to make *proteins that pass along *food chains. Denitrifying bacteria release nitrogen as they *decompose dead organisms and waste.

nitrogen fixation Combining unreactive nitrogen from the atmosphere with other elements to make reactive compounds. Lightning, the *Haber process, and some bacteria in the soil carry out nitrogen fixation. (see *nitrogen cycle)

noble gas A colourless, odourless gas that is a member of group VIII (sometimes called group 0) in the *periodic table of *elements. The noble gases include *helium, *neon, *argon, *krypton, and *xenon. They all occur in small amounts in air and are usually obtained from the *fractional distillation of *liquid air. The noble gases are very unreactive because they have full outer electron *shells.

non-metal A substance whose typical *physical properties in the solid state include brittleness and an in-

normal distribution

206

ability to conduct electricity. Non-metal elements are
included on the right-hand side of the *periodic table.
Typical *chemical properties include the formation of
acidic *oxides and *anions.

normal distribution The range of values to be ex-
pected when measuring a property of a group of objects.
E.g. the heights of maize seedlings planted at the same
time; examination results for a class of science students.
A normal distribution gives a bell-shaped graph. An in-
dividual that does not fit the graph has had its property
altered by a factor not shared by the rest of the group.
(E.g. a maize seedling affected by *mutation; an examination cheat.) ✍

nose The part of a vertebrate concerned with the sense
of smell. Air is drawn through the nose and nasal cavity
by the lungs. The air is warmed and filtered as it passes
through the nasal cavity. Sensors detect chemicals in
the air and send messages to the brain. ✍

nova A star that becomes over a thousand times
brighter over a period of a few days. 10–15 novas are observed each year. A *supernova* results when a star ex-
plodes and its brightness increases by many millions of
times. Six supernovas have been observed over the past
thousand years.

nuclear energy *Energy obtained from *nuclear fission or *nuclear fusion. It is mainly in the form of *thermal (heat) energy. 1 kg of the mixture of uranium
*isotopes used in a nuclear power-station releases the
same amount of energy as burning 25 tonnes of coal. Nuclear energy is used in nuclear *power-stations to generate *electrical energy. (see *nuclear reactor)

nuclear fission The splitting of an atomic nucleus. It
is accompanied by *nuclear radiation and the release of
large amounts of *energy. Nuclear fission can happen
naturally or when a nucleus is struck by a *neutron.
(see *radioisotope)

Normal distribution

Exam results for a class of 25 pupils

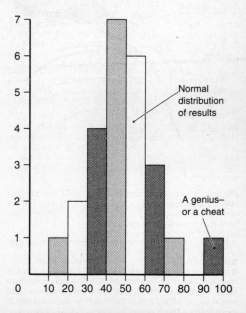

Normal distribution of results

A genius– or a cheat

Nose

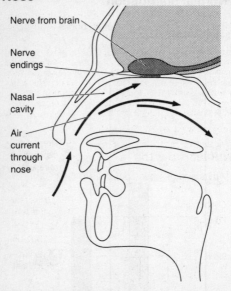

Nerve from brain

Nerve endings

Nasal cavity

Air current through nose

---

nuclear fusion (fusion, thermonuclear fusion) Forming an atomic nucleus by joining together two smaller nuclei of lower *mass number. Nuclear fusion releases large amounts of energy but the temperature must be above 10 000 000 K for it to happen. The Sun's energy is generated by the fusion of hydrogen to make helium.

nuclear radiation *Alpha, *beta, or *gamma radiation given off during the *decay of a *radioisotope. It can damage or destroy living cells. (see *nuclear fission)

nuclear reaction A *reaction that causes a change to the *nucleus of an atom. Nuclear reactions include *nuclear fission and *nuclear fusion.

nuclear reactor A device that uses *nuclear fission to produce energy or *radioisotopes. The core of a nuclear *power-station reactor usually contains *uranium fuel, control rods, and a moderator. The moderator slows neutrons so that they cause nuclear fission in the fuel. Control rods absorb neutrons and so control the *chain reaction and the amount of energy produced.

nuclear waste see *radioactive waste

Nuclear reactor

Advanced gas-cooled reactor (AGR)

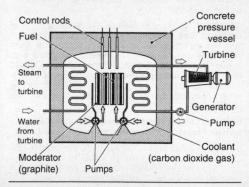

nucleic acid A type of complex *organic molecule found in the nucleus of a living *cell. It consists of a long chain of alternating *sugar and *phosphate molecules. *Bases are attached to each sugar molecule. There are two types of nucleic acid: *DNA and *RNA.

nucleon A particle found in the nucleus of an atom. *Protons and *neutrons are nucleons.

nucleon number Another name for *mass number.

nucleus (pl. nuclei) 1. The central part of an *atom, containing most of the mass of the atom. All nuclei contain one or more *protons. All nuclei except the hydrogen nucleus contain *neutrons. 2. Part of most types of living *cell, made up from *chromosomes enclosed by a membrane. The nucleus controls how a cell works.

nutrient Any substance which a living organism takes in and which helps it to stay alive. Plant nutrients include nitrate, potassium and phosphate *salts from the soil, and carbon dioxide gas. Animal nutrients include *carbohydrates, *proteins, *fats, dietary minerals, *vitamins, and water.

nutrition In living organisms, the whole process of taking in food substances and using them. Nutrition provides the energy and materials an organism needs to stay alive and healthy. Autotrophic organisms (e.g. plants) use *photosynthesis to make their own food from simple substances. Heterotrophic organisms (e.g. animals, fungi) take in complex substances from the other organisms they eat or absorb. (see *nutrient *metabolism)

nylon A thermoplastic man-made *fibre used to make textiles, carpets, and rope. (see *plastic)

nymph A stage in the *life cycle of some insects (e.g. earwig, grasshopper). A nymph looks like the adult, but the wings and reproductive organs are not developed. It develops into the adult without becoming a *pupa.

O

objective lens The *lens(es) in an optical instrument (e.g. microscope) closest to the object.

octane see *alkane

oesophagus (gullet) The tube that carries food from the throat to the *stomach. (see *alimentary canal)

oestrogen A *hormone made in the *ovaries of female mammals. It gives adult females their secondary sexual characteristics. In humans, oestrogen causes adult females to have breasts and softer skin than males. It also partly controls the *menstrual cycle.

ohm (symbol Ω) The *SI unit of electrical *resistance. If a conductor has a voltage of 1 *volt across its ends and a battery makes a current of 1 amp flow, then the conductor has a resistance of 1 Ω. (see *Ohm's law)

Ohm's law A *law stating that the *electric current flowing through a conductor is proportional to the *voltage applied across its ends (provided the temperature remains constant):

$$\frac{\text{voltage (volts)}}{\text{current (amps)}} = \text{constant.}$$

The constant is the *resistance of the conductor, measured in *ohms.

oil A slippery liquid substance that does not dissolve in water. Natural oils are *fats that are liquids at room temperature. Mineral oils are manufactured from *crude oil and are *hydrocarbons with between

Oil refinery

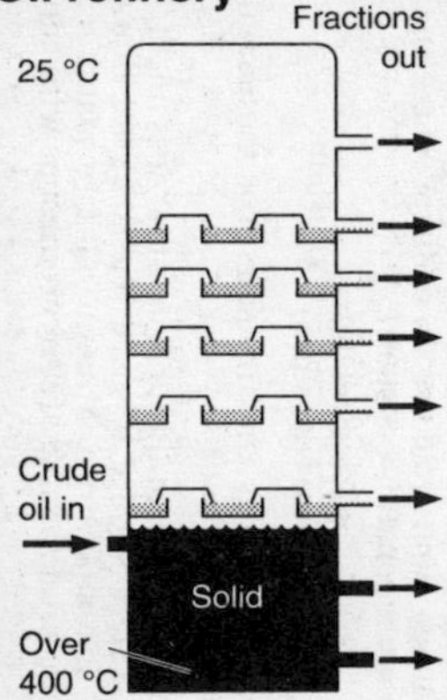

Fraction	Length of carbon chain	Applications
Gas	C_1 to C_4	Separated into methane, ethane, propane and butane – all fuels. Methane is used to make hydrogen.
Petrol	C_4 to C_{10}	Fuel for cars.
Kerosene	C_{10} to C_{16}	Jet fuel. Detergents. Chemicals.
Diesel oil	C_{16} to C_{20}	Fuel for engines. Some cracked.
Lubricating oil	C_{20} to C_{30}	Oil for cars and other machines. Some cracked.
Fuel oil	C_{30} to C_{40}	Fuel for power stations and ships. Some cracked.
Paraffin waxes	C_{40} to C_{50}	Candles, polish, wax-papers, waterproofing, grease.
Bitumen	C_{50} upwards	Pitch for roads and roofs.

Fractions out

25 °C

Crude oil in

Over 400 °C

Solid

20 and 30 carbon atoms per molecule. They are used to lubricate machines.

oil refinery The place where useful substances are made from *crude oil. *Fractional distillation gives refinery gas (containing *methane, *ethane, *propane, and *butane), as well as gasoline (*petrol), *kerosene, *diesel oil, lubricating *oil, *fuel oil, *wax, and *bitumen.   ✍

olefine The old-fashioned name for an *alkene.

omnivore A type of animal that eats both meat and plants. Pigs and humans are omnivores.

operational amplifier (op-amp) An *amplifier, usually made up from *integrated circuits, that has one output and two input connections. The output is – when the inverting input is + and vice versa. The gain of op-amps can be very high; they are most useful when the gain is controlled by negative *feedback. Op-amps are used in audio and TV equipment, measuring instruments, and some types of *computer.

optical fibre A flexible thread of glass along which light travels by a series of *total internal reflections. Bundles of optical fibres make up a 'light pipe' which can be used to look into body cavities such as the stomach, lungs, and uterus. Optical fibres also carry telephone and TV signals, with greater efficiency than a wire the same size.

optic nerve see *eye

orbit 1. In astronomy, the path through space taken by one object as it moves around another. E.g. the Moon has an almost circular orbit around the Earth. 2. The path of an *electron as it travels around the *nucleus in an *atom.

order A group of living organisms with similar *characteristics. Several orders make up a *class.

Lions are in an order called the *carnivores, which also includes tigers, dogs, weasels, and walruses. (see *classification)

ore A type of rock containing a *mineral that can be used in the production of a metal. E.g. *bauxite is an ore of aluminium. (see *gangue)

organ A part of an organism which has its own particular function. *Eyes, *ears, *stomach, and *liver are examples of organs found in animals. Plant organs include *leaves, *roots, and *flowers.

organelle A structure with a special function inside a living *cell. Examples are the *nucleus, *mitochondria, and *ribosomes.

organic compound Originally defined as a compound made in a living *organism; now defined as any *covalent compound containing *carbon. There are about 3 million known organic compounds. This large number is due to the ability of carbon atoms to make four strong bonds and to form into chains and rings. Living organisms and *fossil fuels are made from organic compounds.

organism An animal or plant or anything that lives. All organisms have life processes that include *respiration, *excretion, *reproduction, and the ability to respond to a *stimulus. An organism may consist of a single *cell (e.g. a *bacterium) or many millions of cells (e.g. a tree, a human being).

oscillate To move steadily back and forth between two fixed points. A pendulum oscillates as it swings. The prongs of a tuning-fork and the air in a trumpet oscillate as they make a sound. (see *oscillator)

oscillator An electronic circuit producing an output voltage that *oscillates at a known *frequency. Oscillators control radio transmitters and are used to produce the sound from electronic organs and synthesizers.

oscilloscope An instrument that displays a changing electric signal as a picture on the screen of a *cathode-ray tube. The time base makes a spot of light go back and forth across the screen. The signal is connected to the Y-input and the charge on the Y-plates deflects the spot of light up and down. An oscilloscope can be used to measure the *frequency and *amplitude of a signal.

osmoregulation The processes used by plants and animals to control the composition of fluids inside them. This allows the *cells in the animal to function properly. Osmoregulation in mammals is carried out mainly by the *kidneys. (see *homoeostasis *osmosis)

osmosis A process that happens when two *solutions are separated by a *semi-permeable membrane. More solvent *diffuses through the membrane from the weak solution to the strong solution than in the opposite direction. This is because there is a greater proportion of solvent in the weak solution. ✍

osmotic pressure The *pressure that must be applied to stop *osmosis happening. It is a measure of the concentration of the *solution contained by the *semi-permeable membrane. ✍

ossicle see *ear

ovary An organ which produces female *gametes. In female animals, the ovary produces egg cells (see *ovum). In female humans, it also produces the *hormones *oestrogen and *progesterone. In *flowers, the ovary contains ova which grow into *seeds after *fertilization.

oviduct see *Fallopian tube

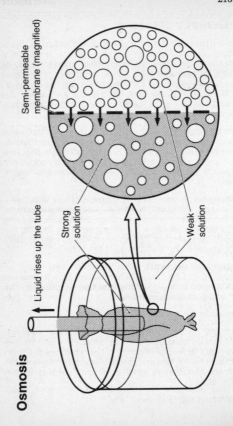

Semi-permeable membrane (magnified)

Strong solution

Weak solution

Liquid rises up the tube

Osmosis

Osmotic pressure

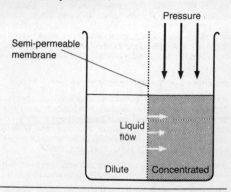

ovulation The release of an *ovum (egg cell) from an ovary. In mammals, the ovum enters the *Fallopian tube and travels down it into the *uterus.

ovule see *flower

ovum (pl. ova) An unfertilized *gamete (egg cell) in a plant or female animal. Ova are made in the *ovaries.

oxidant see *oxidizing agent

oxidation A *chemical reaction where oxygen joins with a substance. E.g. copper reacting with oxygen to form copper oxide. The copper has also lost electrons to the oxygen: $2Cu(s)+O_2(g) \rightarrow 2Cu^{2+}O^{2-}(s)$. Oxidation is therefore said to happen to any atom or group of atoms that loses electrons during a chemical reaction.

oxide A type of *compound that contains oxygen combined with another element. Basic oxides are oxides of metals, e.g. copper oxide CuO. They are solids that react with *acids to form salts. Acidic oxides are oxides of non-metals, e.g. sulphur dioxide SO_2. They are usually gases and react with *bases to form salts. Amphoteric oxides have the properties of both acidic and basic oxides, e.g. aluminium and zinc oxides.

oxidizer see *oxidizing agent

oxidizing agent (oxidizer, oxidant) A chemical *reagent that causes *oxidation to happen to an atom or group of atoms by removing electrons from it. Oxidizing agents include oxygen, chlorine, and iron(III) ions (Fe^{3+}), and *oxo-ions such as dichromate $Cr_2O_7^{2-}$ and manganate(VII) MnO_4^- (permanganate).

oxo-ion An ion containing oxygen together with other atoms, e.g. sulphate SO_4^{2-}, dichromate $Cr_2O_7^{2-}$.

oxygen (symbol O) A colourless gaseous *element that is a member of group VI of the *periodic table. It makes up 21% of *air and is very reactive. Oxygen exists as molecules O_2. It is extracted by the *fractional distillation of *liquid air and is used in making *steel, for high-temperature cutting and welding flames, and in breathing apparatus. Most living organisms use oxygen during *respiration. Oxygen has a second *allotrope, *ozone.

oxygen debt The amount of oxygen required to remove the products of *anaerobic respiration. Running very fast requires more energy than muscles can produce from *aerobic respiration alone. Extra energy comes from anaerobic respiration which does not use oxygen, but which produces lactic acid. Heavy breathing continues after running stops to break down the lactic acid and repay the oxygen debt.

oxygen lance A pipe that sends a blast of oxygen into molten *pig iron during the manufacture of *steel.

ozone (trioxygen) A colourless and poisonous gas that has high *reactivity. Its formula is O_3. Ozone is formed when *ultraviolet radiation or an electric spark passes through oxygen. (see *ozone layer)

ozone layer A layer of *ozone 15–50 km above the Earth's surface, caused by sunlight acting on oxygen. The ozone layer absorbs harmful *ultraviolet radiation and protects organisms on Earth. Human activity releases *chlorofluorocarbons that are removing ozone from the layer.

P

pancreas In vertebrates, a *gland that makes digestive *enzymes and passes them into the small *intestine. Part of the pancreas also works as an *endocrine gland, producing the *hormones *insulin and *glucagon.

paraffin 1. (kerosene) A liquid containing *hydrocarbons with 11 or 12 carbon atoms. It is made by distilling *crude oil and is used as a fuel for jet engines and for heating. 2. The old-fashioned name for an *alkane.

parallel connection A method of connecting together separate components in an electric *circuit. The *electric current divides, flows through the components, and then combines again. Two bulbs connected in parallel each receive the full voltage of the battery and so shine equally brightly. Two similar bulbs take twice the current of a single bulb. Two 6-*volt batteries connected in parallel have a *voltage of 6 volts but together they can supply twice the current of a single 6-volt battery. (see *series connection) ✍

parasite An organism that takes its food from another living organism called the host. Parasites can cause discomfort, disease, or death. Fleas and tapeworms are parasites. Malaria is a disease caused by a parasitic *protozoon that lives in mosquitoes and humans.

particle A tiny piece of *matter. Particles include *atoms and *molecules, *ions, and *alpha and *beta particles. *Electrons, *protons, and *neutrons are

Parallel connection

Bulbs in parallel

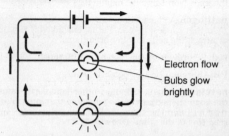

Electron flow

Bulbs glow brightly

Cells in parallel

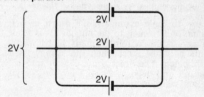

fundamental particles. Large particles include *pollen grains and dust.

pascal (symbol Pa) The *SI unit of *pressure. 1 Pa is equivalent to a *force of 1 newton on an area of 1 square metre.

pasteurization Treating milk to destroy *microbes that may cause disease. The milk is heated to 72 °C for 15 minutes and is then rapidly cooled to below 10 °C.

patella The small round kneecap bone that covers the knee *joint in most mammals. (see *skeleton)

pathogen A *bacterium, *virus, *fungus, or *protozoon that causes disease. Pathogens give off *toxins or attack the cells of other organisms.

p.d. Abbreviation for *potential difference.

PE Abbreviation for *potential energy.

pelvis In humans and many other land vertebrates, the bowl-shaped part of the *skeleton that joins to the rear legs and the spine. In fish, the pelvis joins the rear fins to the spine. (see also *kidney)

penicillin An *antibiotic substance made by a mould called *Penicillium notatum*. It is used to treat infections caused by some types of *bacteria.

penis In male mammals, the organ used to pass *urine and *semen to the outside. It is made from spongy tissue containing many blood vessels. The penis becomes erect when the spongy tissue fills with blood. The erect penis can fit into a female *vagina and deposit semen near the *cervix. (see *sexual reproduction)

pentane see *alkane

pepsin An *enzyme made in the stomach of a vertebrate animal. It helps with the *digestion of *proteins in food.

peptide A group of two or more *amino acids joined together. Dipeptides contain two amino acids; tripeptides contain three. Peptides join to make *polypeptides contained in *proteins. Peptides result during the *digestion of proteins.

223   **peristalsis**

perennial A type of plant that lives for a number of years. Trees and shrubs (e.g. oak and privet) are perennial plants that grow larger year after year. The stems and leaves of *herbaceous perennial plants (e.g. rhubarb and daffodils) die down each autumn. New shoots grow up each spring from the parts of the plant that are underground. (see *annual *biennial)

period 1. A horizontal row of *elements in the *periodic table. Each element has one more *proton and one more *electron than the element to its left. The elements in a period show a trend from metal to non-metal properties as *atomic number increases. 2. The time taken for a moving object (e.g. a rotating wheel or orbiting satellite) to return to its original position. The period of the Earth's orbit is 365.25 days; for a *geostationary satellite, the period is exactly one day. 3. The time taken for one complete oscillation of a vibration or *wave.

Period (seconds) = 1/frequency (hertz).

4. see *menstrual cycle.

periodic table A table that lists the *elements in order of increasing *atomic number. It consists of eight *groups of elements listed in columns labelled I, II, III, . . ., VIII and a block of *transition metals. Group VIII is sometimes labelled group 0. Elements in the same group have similar properties and the same number of electrons in their outer *shells. Rows of elements are called periods and show a steady trend in properties from left to right of metal to non-metal character. (see *alkali metal *alkaline earth metal *halogen *noble gas)   ✍

peristalsis The process that squeezes food along inside the *alimentary canal. Rings of muscle around the *intestines and the *oesophagus contract and relax in turn.

Periodic table

Group

	I	II
1	H ^{1}	
2	Li ^{3}	Be ^{4}
3	Na ^{11}	Mg ^{12}
4	K ^{19}	Ca ^{20}
5	Rb ^{37}	Sr ^{38}
6	Cs ^{55}	Ba ^{56}
7	Fr ^{87}	Ra ^{88}

The transition metals

Sc ^{21}	Ti ^{22}	V ^{23}	Cr ^{24}	Mn ^{25}	Fe ^{26}	Co ^{27}	Ni ^{28}	Cu ^{29}	Zn ^{30}
Y ^{39}	Zr ^{40}	Nb ^{41}	Mo ^{42}	Tc ^{43}	Ru ^{44}	Rh ^{45}	Pd ^{46}	Ag ^{47}	Cd ^{48}
La ^{57}	Hf ^{72}	Ta ^{73}	W ^{74}	Re ^{75}	Os ^{76}	Ir ^{77}	Pt ^{78}	Au ^{79}	Hg ^{80}
Ac ^{89}	Unq ^{104}	Unp ^{105}	Unh ^{106}	Uns ^{107}					

Group

III	IV	V	VI	VII	VIII
					He ^{2}
B ^{5}	C ^{6}	N ^{7}	O ^{8}	F ^{9}	Ne ^{10}
Al ^{13}	Si ^{14}	P ^{15}	S ^{16}	Cl ^{17}	Ar ^{18}
Ga ^{31}	Ge ^{32}	As ^{33}	Se ^{34}	Br ^{35}	Kr ^{36}
In ^{49}	Sn ^{50}	Sb ^{51}	Te ^{52}	I ^{53}	Xe ^{54}
Tl ^{81}	Pb ^{82}	Bi ^{83}	Po ^{84}	At ^{85}	Rn ^{86}

Ce ^{58}	Pr ^{59}	Nd ^{60}	Pm ^{61}	Sm ^{62}	Eu ^{63}	Gd ^{64}	Tb ^{65}	Dy ^{66}	Ho ^{67}	Er ^{68}	Tm ^{69}	Yb ^{70}	Lu ^{71}
Th ^{90}	Pa ^{91}	U ^{92}	Np ^{93}	Pu ^{94}	Am ^{95}	Cm ^{96}	Bk ^{97}	Cf ^{98}	Es ^{99}	Fm ^{100}	Md ^{101}	No ^{102}	Lr ^{103}

permanent magnet A *magnet that is constantly surrounded by a steady *magnetic field and that does not rely on an electric current to maintain its magnetism. A permanent magnet loses its magnetism when heated or beaten.

permeable Describing a solid (usually *porous) that allows a *fluid to pass through. E.g. *sandstone and filter paper are permeable to water and to air.

peroxide A type of compound that contains two oxygen molecules joined together, e.g. hydrogen peroxide H–O–O–H, H_2O_2. Most peroxides decompose easily, giving off oxygen gas.

Perspex (polymethylmethacrylate) A hard, transparent, brittle thermoplastic. It can be used instead of glass. (see *plastic)

petrol (gasoline) A *volatile liquid containing *hydrocarbons with 5 to 8 carbon atoms per molecule. It is made from *crude oil and is used as a fuel in petrol engines. Petrol bought at a garage has many additives to improve engine performance.

petrol engine see *internal combustion engine

petroleum see *crude oil

pH A scale of numbers between 0 and 14 that shows the strength of an *acid or an *alkali. Acidic solutions have a pH less than 7; alkaline solutions have a pH greater than 7. The pH of a solution can be found by adding universal *indicator and comparing the resulting colour with a chart.

phagocyte A type of white *blood cell that helps to defend an animal against *microbes that cause diseases. Phagocytes engulf and digest invading microbes and other foreign particles.

phalanges (sing. phalanx) In humans, the bones that make up the fingers and toes. (see *skeleton)

phenotype A *characteristic of an organism that can be seen, such as brown eyes, a yellow flower colour, or curly hair. Each phenotype is caused by a set of *genes called the *genotype.

phloem In many types of plant, a *tissue made up from tubes. Phloem is part of the *vascular system of a plant. It carries dissolved *glucose and other substances from the leaves where it was made by *photosynthesis to places where it is needed. (see *xylem)

phosphate An *ion and an acid *radical with the formula PO_4^{3-}. Phosphates form *salts, e.g. trisodium phosphate Na_3PO_4. *Alkali metal phosphate salts are soluble in water. All others are insoluble.

phosphorus (symbol P) A soft and highly reactive *non-metal *element that is a member of group V of the *periodic table. It exists as several *allotropes; red phosphorus is used to make matches, and white phosphorus (stored under water) inflames when exposed to air.

photocell (photoelectric cell) A device that gives out an electric signal when light or other *electromagnetic radiation falls upon it. A photocell can be used to measure the intensity of light.

photosynthesis The process used by *plants to make their own food. Photosynthesis takes place inside the *chloroplasts in plant *cells. *Chlorophyll in the chloroplasts traps energy from light, which helps to make *glucose from carbon dioxide and water. Oxygen is released at the same time. The glucose is then used during *respiration or is changed into *starch and stored.

phylum (pl. phyla) A group of animals with similar *characteristics. Several phyla make up a *kingdom. Plants are grouped in *divisions, not phyla. Lions are in a phylum called the *vertebrates, which includes all animals with a backbone. (see *classification)

physical change A change happening to a substance that does not alter the *chemical properties of the substance. A physical change is usually easy to reverse. Physical changes happen when substances expand or contract, *melt, *boil, *freeze, *sublime, *crystallize, *condense, or *dissolve.

physical property A description of the appearance of a substance or of a *physical change that can happen to it. E.g. two physical properties of water are: liquid at room temperature; boiling-point = 100 °C at *standard pressure.

physics The study of how matter and energy affect each other.

pig iron An impure form of iron that is made in a *blast-furnace. Pig iron contains up to 4% carbon and is very hard and brittle. Some is used as cast iron to make the casings of engines, but most is converted into *steel.

pigment An *insoluble coloured substance. E.g. paint contains particles of pigment mixed in with a liquid; the choroid in the human *eye contains pigment that absorbs light.

pinna see *ear

pipette A piece of apparatus used to measure out accurately a fixed volume of liquid. It is filled to the calibration mark by suction and the liquid is then allowed to run out. Pipettes are made in standard sizes of 5, 10, 20, 25, 50 cm³.   ✍

piston A part in some *machines and *engines that slides up and down inside a cylinder. The moving piston in a pump or syringe draws fluid into the cylinder or forces it out. Burning gases in petrol and diesel engines expand and push the piston along the cylinder. (see *internal combustion engine)

pitch 1. How different frequencies sound to the ear. High frequencies (e.g. from a football referee's whistle) have a high pitch; low frequencies (e.g. the bass notes of a piano) have a low pitch. 2. A brown/black sticky solid that is the residue when *crude oil or coal *tar is distilled. It is used in making road surfaces, to waterproof roofs, and as a fuel.

pituitary gland An *endocrine gland attached to the underneath of the brain. It makes many *hormones, including one which controls growth.

Pipette

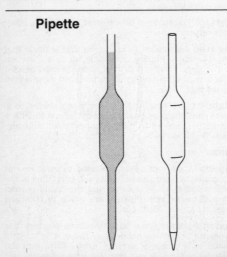

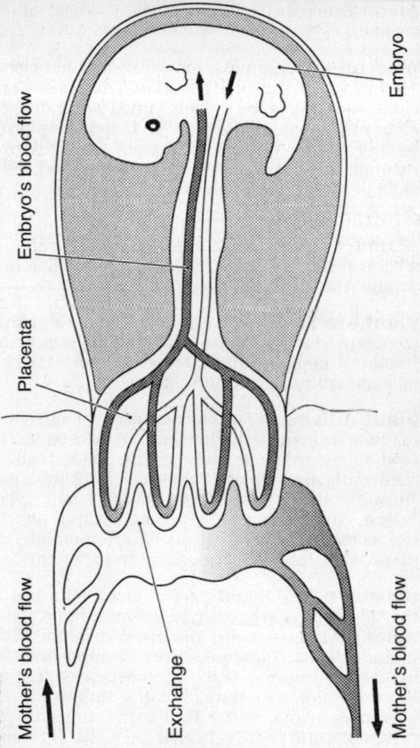

Placenta

Mother's blood flow

Embryo's blood flow

Embryo

Placenta

Exchange

Mother's blood flow

pivot The point around which a wheel or *lever turns.

placenta In mammals, the organ that nourishes an *embryo growing in its mother's *uterus. The placenta contains blood vessels joined separately to the embryo and to the mother. The blood of embryo and mother do not mix. Nourishment and oxygen pass through the *umbilical cord from the mother to the embryo in exchange for carbon dioxide and other waste substances. ✍

planet A large body that moves in an *orbit around a central star. E.g. Earth, Mars, Pluto, which revolve around the Sun. (see *solar system)

plankton Tiny plants (phytoplankton) and animals (zooplankton) that float about in the sea or lakes. Plankton include *algae, *protozoa, and the *larvae of some animals. (see *producer)

plant A living organism that uses *photosynthesis to make its own food. Plants do not have to search for food so they do not move about. Plant *cells have *cell walls and contain *chloroplasts. Plants include *mosses and *liverworts, *ferns, some types of *algae, *angiosperms (flowering plants), and *gymnosperms (conifers). All plants are members of the plant *kingdom. (see *bud *leaf *root *stem) ✍

plasma 1. The liquid part of blood obtained when the *blood cells are removed. Human blood plasma is water containing many dissolved substances. These include *salts, *glucose, *urea, *amino acids, *hormones, *vitamins, and the substances that cause blood to *clot. 2. A state of matter that exists at temperatures above 50000 K. Plasma consists of free *electrons and positively charged *ions. It is found in *nuclear fusion reactions (e.g. the Sun's atmosphere) and in *discharge lamps.

Plant

A flowering plant

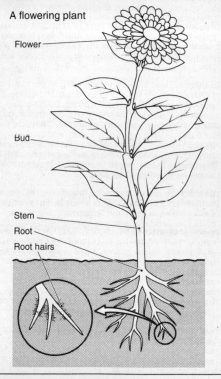

Flower

Bud

Stem
Root
Root hairs

plasmolysis The cause of leaves wilting if they lose too much water. The *protoplasm shrinks away from the *cell wall as a result of plasmolysis.

plaster A powder made by roasting and grinding *gypsum with some *limestone and *clay. It sets into a hard solid when mixed with water. A coat of plaster gives a smooth flat surface to walls inside buildings.

plastic 1. Describing a substance that permanently changes its shape when squeezed, e.g. Plasticine. 2. A *polymer molecule manufactured from crude oil, commonly called a plastic. Plastics contain up to 50000 *monomer units per molecule. They are good electrical insulators, do not rot, and are easy to shape in moulds. Thermoplastics soften when heated and harden when cooled (e.g. *polythene, *PVC). Thermosets harden when heated (e.g. Melamine) and cannot be re-softened by heat.

platelet A small particle found in the *blood. Platelets help blood to clot in wounds. They are parts of cells found in red bone *marrow.

plate tectonics The theory that the outer solid layer (*crust) of the Earth is made up from separate solid pieces called plates. Plate tectonics suggests that there are six major plates which move relative to each other (speeds are approximately 3 cm/year), causing *continental drift. Mountain formation, earthquakes, and volcanoes are explained as changes happening at the edges of plates as they collide, separate, or rub past each other.

platinum (symbol Pt) A silvery-white *transition metal *element that is very unreactive. Platinum is extracted from impurities that separate during *copper refining. It is used in jewellery and alloys and for *electrodes in laboratory apparatus. It is used as a

*catalyst during the manufacture of *nitric acid from *ammonia.

plumule Part of the *embryo in a *seed which becomes a *shoot. (see *germination)

plutonium (symbol Pu) A silvery poisonous metal that is very *radioactive. Plutonium is made in *nuclear reactors when *uranium-238 nuclei absorb neutrons. The most important isotope is ^{239}Pu, which is used (like uranium-235) in atomic weapons and in some nuclear reactors.

pneumatic 1. Describing a *machine that moves mechanical energy from one place to another by means of compressed air. A pneumatic road drill contains a piston driven up and down by high-pressure air from a compressor. 2. Describing an object containing compressed air, e.g. a pneumatic car tyre.

poikilotherm (cold-blooded animal) An animal whose internal body temperature is the same as the external environment. Poikilotherms such as *reptiles and *arthropods become sluggish in cold weather. All animals except birds and mammals are poikilotherms.

polarization The process of producing polarized *electromagnetic radiation, made up from *waves whose electrical (or magnetic) *field vibrates in one plane (direction) only. Polarized light results from passing rays of ordinary (randomly polarized) light through a *Polaroid filter. The direction of polarization of radio waves depends on the arrangement of the transmitting *aerial. E.g. a vertical aerial transmits vertically polarized waves.

Polaroid A type of filter that allows light *waves vibrating in one plane (direction) only to pass through. (see *polarization)

pole see *magnetic pole

pollen The fine dustlike grains discharged from the male part of a *flower. (see *pollination)

pollen tube see *flower

pollination The transfer of pollen from the male to the female reproductive organs in flowering plants. This process leads on to *fertilization and *seed production. In flowering plants (*angiosperms), pollen is transferred from a male anther in a *flower to a female stigma in a *flower. Self-pollination takes place within the same flower. Cross-pollination takes place when pollen is transferred from one flower to another flower of the same *species. Cross-pollination is carried out by insects (insect pollination) or by the wind (wind pollination). Flowers with bright petals, strong scents, and *nectaries attract insects. Insects feed from the nectaries and carry pollen from one flower to another on their bodies. Grasses do not have bright petals, and *gymnosperms (e.g. pine and fir trees) have cones instead of flowers. These plants depend on wind to carry the pollen.

pollution A change for the worse in the natural environment caused by human or natural activity. The main types of pollution and their causes are: air pollution (fumes from vehicle exhausts and power-stations, ash from volcanoes); water pollution (factory waste, farm fertilizers, *herbicides, *insecticides, and untreated sewage); noise pollution (aircraft, traffic); land pollution (litter, dumping).

polyester A thermoplastic man-made fibre used to make textiles (e.g. Terylene). Polycotton is a mixture of cotton and polyester *fibres. (see *plastic)

polymer A substance containing chain-like molecules made by joining *monomer molecules end to end. Natural polymers include *proteins and *polysaccharides (e.g. *starch). Man-made polymers include *plastics such as *polythene and *PVC.

235

Polymerization

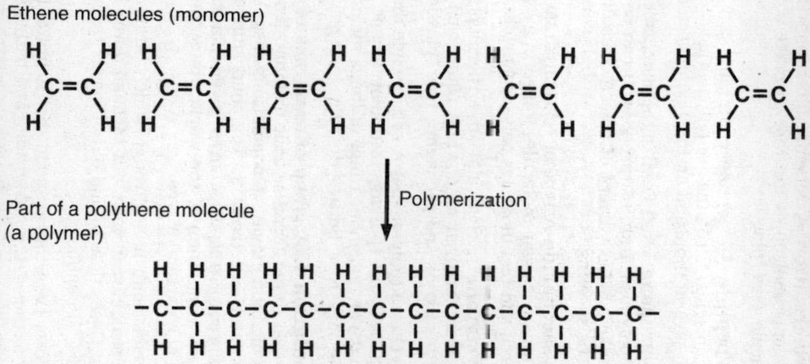

Ethene molecules (monomer)

→ Polymerization →

Part of a polythene molecule (a polymer)

polymerization A *chemical reaction that joins *monomer molecules together to make a *polymer molecule. (see *plastic) ✎

polypeptide A *peptide containing ten or more *amino acids. *Proteins are polypeptides containing up to several thousand amino acid molecules.

polypropene A soft flexible thermoplastic made by the *polymerization of *propene. It can be moulded to make a wide variety of objects including tough *plastic sheeting.

polysaccharide A type of *carbohydrate made up from many *sugar molecules joined in a long chain. *Starch and *cellulose are polysaccharides.

polystyrene A clear glass-like thermoplastic usually manufactured as a white rigid foam used for heat insulation and packaging. (see *plastic)

polythene (polyethene) A soft flexible thermoplastic made by the *polymerization of *ethene. It is used to make an enormous range of items including polythene bags, toys, bottles, dustbins, and bowls.

population 1. A group of organisms of the same *species living together in a *community. A woodland *habitat may have populations of squirrels, spiders, and oak trees. 2. The total number of the members of a species in a given place. E.g. the human population of the world; the spider population of a particular tree.

pore A small hole through which liquids or gases can pass. Sweat flows out through pores in human skin. Gases flow in and out of leaves through pores called *stomata.

porous Describing a solid that is full of tiny holes and passages. E.g. *sandstone and bath sponges are porous.

potassium (symbol K) A soft silvery *metal *element that is a member of group I in the *periodic table (the *alkali metals). Potassium is extracted by *electrolysis of molten potassium chloride. It is a highly reactive element. Many of its salts have important industrial uses. Potassium is an *essential element for living organisms.

potential difference (p.d.) Another name for *voltage, a measure of the work done by a unit of *charge as it flows between two points in an electrical circuit. (see *volt)

potential energy (PE) *Energy that is stored in an object or substance due to its state or position. E.g. a mass held above the ground, a stretched spring, and a compressed gas all have potential energy. It is converted to *kinetic energy when the object or substance is released and allowed to return to its normal state.

potentiometer An electrical *resistor fitted with a sliding contact. A potentiometer is used as a voltage divider. It passes on only a part of the voltage applied across its ends, depending on the position of the sliding contact.

potometer A device used to measure the rate of water loss of a plant shoot due to *transpiration. The cut shoot draws water from a reservoir and the rate of transpiration is shown by the movement of an air bubble against a scale.   ✍

power The rate at which work is done or energy is changed into another form. The *SI unit of power is the *watt (W).

$$\text{Power} = \frac{\text{work done or energy demand}}{\text{time taken}}.$$

In an electrical circuit, power = volts×amps.

Potometer

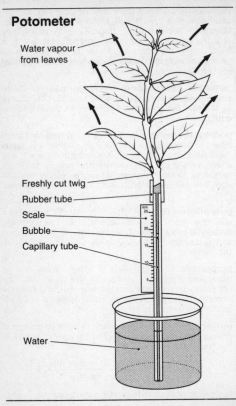

Water vapour
from leaves

Freshly cut twig

Rubber tube

Scale

Bubble

Capillary tube

Water

power-station A place where *electrical energy is generated, usually from burning *fossil fuels or from *nuclear energy. Heat is used to boil water to make steam that drives a *turbine. The turbine powers an *alternator. Cold water from a cooling tower condenses the used steam to water which returns to the boiler. (see *hydroelectricity)

precipitate An insoluble solid made by the reaction of two solutions. E.g. mixing solutions of sodium chloride and silver nitrate forms a precipitate of white silver chloride:

$$NaCl(aq)+AgNO_3(aq) \rightarrow AgCl(s)+NaNO_3(aq).$$

predator An animal which hunts other animals for food. Predators are carnivores and are secondary or tertiary *consumers (see *food chain). Examples include tigers, moles, and spiders.

premolar A type of *tooth used for grinding food. Premolars are wide and ridged but are smaller than *molar teeth. They are found in front of the molar teeth and usually have one or two roots.

pressure The force pressing on 1 m^2 of a surface.

$$\text{Pressure} = \frac{\text{force}}{\text{area}}.$$

Units of pressure include newtons per square metre (N/m^2) and pascals (Pa). 1 Pa is the same as 1 N/m^2.

primary cell A type of electrochemical *cell (e.g. a *zinc–carbon cell) which cannot be recharged when exhausted (flat).

primary colour One of three coloured lights (e.g. red, green, and blue) that give white light when mixed together in equal amounts. Any colour can be made by mixing together different amounts of primary colours.

Prism

Total internal reflection in a prism

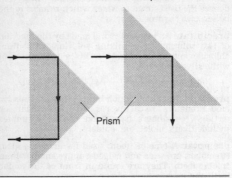

Prism

---

printer A machine attached to the output of a *computer that writes words and numbers on to paper. Some printers can also draw diagrams.

prism A block of transparent material, usually triangular in shape, that can be used for the *reflection or *dispersion of light. Light *rays striking a prism at the correct angle will undergo *total internal reflection. White light can pass through a prism and be dispersed into a *spectrum of colours. ✍

producer An organism that provides food and energy (by *photosynthesis) for other organisms in a *food chain. *Plants are the producers on the land. Phytoplankton are the producers in the sea. (see *plankton)

product A substance that is a result of a *chemical reaction.

progesterone In human females, a *hormone made in the *ovaries. It partly controls the *menstrual cycle.

program A set of instructions that makes a *computer carry out a particular task. E.g. depending on its program, a computer may be a word processor, a robot controller, or a games machine. The program controls how the flow of *data through a computer is processed.

progressive wave A *wave that travels from one place (the source) to another (the destination). (compare *stationary wave)

propagation 1. Increasing the numbers and spreading the members of a *species of living organism. E.g. propagation of a plant, animal, disease. 2. The movement of a wave away from its source E.g. the propagation of sound through air and of light through a vacuum.

propane A gas that is a *hydrocarbon and an *alkane. Its formula is C_3H_8. Propane is made by *fractional distillation of *crude oil. It liquefies easily when compressed and is used as a fuel ('bottled gas').

propanol see *alcohol

propene A gas that is an *unsaturated *hydrocarbon and an *alkene. Its formula is C_3H_6. Propene is made by *cracking hydrocarbons from *crude oil. It is used to make the *plastic polypropene.

property Something about an object or substance that can be observed or measured. E.g. two properties of water are that it (1) is wet and (2) boils at 100 °C.

propylene The old-fashioned name for *propene.

prostate gland A *gland connected to the *urethra in male mammals. A liquid from the prostate gland mixes with sperms to make *semen. (see *sexual reproduction)

protein A type of substance found in all living organisms. Proteins are made up from long chains called *polypeptides which contain up to several thousand *amino acid molecules. Examples include *haemoglobin in blood, albumin in egg-white, the *hormone *insulin, *muscle fibres, and *enzymes. Animals take in proteins in their food. These body-building foods include meat, fish, peas, and beans.

protist A type of very simple organism that makes up a *kingdom in some forms of *classification. Protists include *algae, *bacteria, *fungi, and *protozoa. These organisms are usually grouped into the *protoctista, *monera, and fungi kingdoms.

protium (normal hydrogen) The most common *isotope of *hydrogen, containing one proton only in the nucleus.

protoctista A group of very simple living organisms. It includes *protozoa and some types of *algae. All protoctista are members of the protoctista *kingdom.

proton A particle contained in the *nucleus of all *atoms. It has a positive *charge equal in value to the negative charge on an *electron. The mass of a proton is about 1840 times the mass of an electron.

proton number Another name for *atomic number.

protoplasm The total contents of a living *cell. The *cytoplasm and the *nucleus together make up the protoplasm.

protozoon (pl. protozoa) Any *protist or simple animal that consists of a single cell. Most protozoa are *saprophytes. Some are *parasites and can cause diseases, e.g. the *Plasmodium* protozoon causes malaria.

pseudopodium (pl. pseudopodia) A finger-shaped part pushed out by a *unicellular organism. Two pseudopodia are used to trap food. Some *protozoa (e.g. *Amoeba*) move by repeatedly pushing out a pseudopodium and then flowing into it.

PTFE (polytetrafluoroethene) A soft slippery thermoplastic used to coat non-stick cooking utensils. It is also used to lubricate surfaces that slide over each other (see *plastic)

puberty A stage in the growth of a human being. During puberty, a child becomes a young adult and is capable of reproduction. Puberty begins when the *gonads start to produce sex *hormones.

pulley A simple *machine made of a rope passing over a wheel. An *effort applied at one end of the rope moves a *load attached to the other. A pulley changes the direction of the effort. Two or more pulleys working together decrease the effort needed to move the load. ✍

pulmonary artery see *heart

pulmonary vein see *heart

pulp cavity see *tooth

pulsar A *neutron star that sends out pulses of radiation every 0.03 to 4 seconds.

pumped storage Using electricity that is unwanted during the night to pump water up a hill and into a reservoir. During daytime periods of high demand, the water is used to generate *hydroelectricity.

Pulley

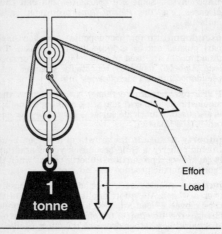

Effort

Load

pupa One non-feeding stage in the *life cycle of some insects and other *invertebrates. *Metamorphosis changes a *larva into a pupa. A chrysalis is the pupa of a butterfly. (see *nymph)

pupil see *eye

pure A pure substance contains one *element or one *compound only. A pure mixture contains two or more pure substances with no trace of any additional substances.

Pyramid of numbers

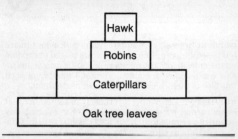

| Hawk |
| Robins |
| Caterpillars |
| Oak tree leaves |

PVC (polyvinylchloride) A thermoplastic used to make tarpaulins, pipes, and the insulation in electric cables. (see *plastic)

pyramid see *kidney

pyramid of mass A diagram that shows the total mass (*biomass) of the organisms in each of the *trophic levels in a *food chain. (see *pyramid of numbers)

pyramid of numbers A diagram that shows the numbers of organisms in each of the *trophic levels in a *food chain. The first trophic level (plants) usually contains the largest number and the highest trophic level (large carnivores) the smallest number.

pyrites A *mineral containing iron sulphide FeS.

pyrogallol A solid that is dissolved in a concentrated solution of sodium hydroxide and used to remove oxygen from air.

Q

quadrat In *ecology, a square frame with sides 1 metre long used in *sampling. Quadrats are placed at random on the ground and the numbers of chosen organisms inside each square are noted. The total number of each *species is then calculated for the whole *habitat.

qualitative Describing a *property of something that is concerned with what it is, but not measuring any quantities related to it. E.g. a food chemist uses qualitative analysis to name the *food additives in a meat pie. (see *quantitative)

quantitative Describing something that is concerned with the measurement of an amount. E.g. a food chemist uses quantitative analysis to find out how much vitamin C is in a bottle of orange juice. (see *qualitative)

quantum theory A theory that describes how *matter and *radiation affect each other. Quantum theory suggests that electrons orbiting in an atom can only have certain fixed *energy levels (quantum levels). They cannot have energies between these levels. The theory accounts for an atom giving off radiation of a certain fixed energy when an electron moves from a higher to a lower energy level. Quantum theory leads to wave mechanics, which explains how particles can have the same properties as *waves.

quark One of a set of fundamental *particles from which all other fundamental particles (e.g. *electrons, *protons, *neutrons) are made.

quartz A common *mineral made up from crystals of silicon dioxide SiO_2. It originates in *igneous rocks such as granite. Slices of man-made quartz are used to control the frequency of electronic *oscillators used in some clocks and radio transmitters.

quasar Abbreviation for quasistellar object. Quasars are the most distant objects in the Universe, having extremely large *red shifts.

quicklime see *lime

R

radar Using radio waves to locate distant objects such as aircraft. An aerial sends out a narrow beam of *microwaves which are reflected by the aircraft. The time taken for the microwaves to return and the direction the aerial is pointing give the distance and position of the aircraft.

radiation 1. Energy travelling in the form of electromagnetic waves (e.g. light, X-rays). (see *electromagnetic radiation) 2. A stream of particles, especially *alpha particles, *beta particles, or *neutrons.

radical A group of atoms (or sometimes a single atom) that forms part of a *compound. Acid radicals are *anions that come from acids.

Acid	acid radical
sulphuric H_2SO_4	sulphate SO_4^{2-}
	hydrogensulphate HSO_4^-
hydrochloric HCl	chloride Cl^-
nitric HNO_3	nitrate NO_3^-
carbonic H_2CO_3	carbonate CO_2^{2-}
	hydrogencarbonate HCO_3^-

Acid radicals form *salts when combined with metal or ammonium ions, e.g. Na_2SO_4, NH_4Cl.

radicle Part of the embryo in a *seed which becomes the *root. (see *germination)

radioactive Describing a substance that contains atoms with unstable *nuclei that can *decay and emit *nuclear radiation. (see *radioisotope)

radioactive waste (nuclear waste) Waste material that contains *radioisotopes. Radioactive waste comes from mining *radioactive *ores, running *nuclear power-stations, and making nuclear weapons. It is dangerous to all living organisms. High-level waste must be safely stored for many thousands of years.

radioisotope An *isotope of an element that is *radioactive. Naturally occurring radioisotopes are most common in elements that have an atomic number greater than 82. Artificial radioisotopes are made by bombarding elements with *nuclear radiation in a *nuclear reactor.

radio waves *Electromagnetic radiation with a *frequency between 3000 Hz (3 kHz) and 3×10^{11} Hz (300 000 MHz). Radio waves are used for broadcasting (radio and television), communications (telephone and *microwave links), and navigation (*radar). (see *ionosphere)

radius One of the two bones that join the wrist to the elbow of humans and many other vertebrates. (see *skeleton)

RAM Abbreviation for *random access memory.

r.a.m. Abbreviation for *relative atomic mass.

random access memory (RAM) A type of memory circuit used in *computers. A RAM is made up from *integrated circuits and contains a temporary store of *programs or *data that can be easily changed. E.g. a RAM contains the information being typed into a word processor.

rate The amount of a change that happens in a unit of time. E.g. the rate that an object moves along its path is its speed, which may be measured in metres per second; the rate that water flows through a pipe may be measured in dm³ (or litres) per second. (see *reaction rate)

raw material Substances obtained from natural *resources and used to make other substances. E.g. iron ore is the raw material used to make iron.

ray 1. A line drawn to show the direction of travel of a *wave. A ray is at a right angles (90°) to the *wavefronts of a wave. 2. A narrow beam of radiation, e.g. gamma rays. (see *gamma radiation)

rayon A silky man-made *fibre used in textiles. It is manufactured from *cellulose extracted from wood.

reactant A substance that is changed into a *product by a *chemical reaction.

reaction 1. see *chemical reaction *nuclear reaction *chain reaction. 2. One of the results of a *force acting on an object. E.g. gravity causes a book to exert a downward force on a table. The table exerts an opposite reaction force upwards on the book. The book does not move because force and reaction are equal in magnitude and opposite in direction. (see *jet propulsion)

Ray

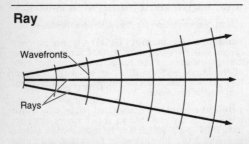

Wavefronts

Rays

Reaction rate

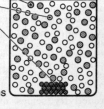

Acid particle
Water molecule
Magnesium atoms

At the start, there are plenty of magnesium atoms and acid particles. But they get used up during successful collisions

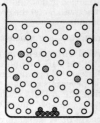

After a time, there are fewer magnesium atoms, and the acid is less concentrated, so the reaction slows down

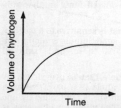

This means that the slope of the reaction curve decreases with time, as shown here

reaction rate The amount of *reactant used up in a *chemical reaction during a unit of time. It is measured in grams per second or *moles per second. High reaction rates result from using high concentrations of reactants and high temperatures, by using a suitable *catalyst, and by powdering solid reactants to give large surface area.   ✍

reactivity The ease and speed with which a substance takes part in *chemical reactions. *Metal reactivity increases with ease of *electron loss. *Non-metal reactivity increases with ease of electron gain. The reactivity of *compounds depends on *stability to heat and the effects of different *reagents.

reactivity series of metals A list of metals arranged in descending order of reactivity. The more reactive a metal the greater the *stability of its *compounds.   ✍

reactor see *nuclear reactor

read only memory (ROM) A type of memory circuit used in *computers. A ROM is made up from *integrated circuits and contains a permanent store of *programs or *data that a computer needs to function. E.g. a ROM changes the letters on a keyboard into *bytes of *binary code.

reagent A substance that can be involved in a *chemical reaction when mixed with another substance.

receptor A type of cell that is sensitive to a particular *stimulus. Receptors send impulses along sensory *nerves. *Eyes, *ears, *nose, and *skin all contain receptors.

recessive allele An *allele that is overridden by the effect of a *dominant allele in a pair. (NB each *characteristic of an organism is controlled by one or more pairs of alleles.) The allele for blue eyes is

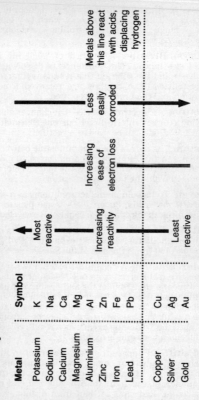

Reactivity series of metals

Metal	Symbol
Potassium	K
Sodium	Na
Calcium	Ca
Magnesium	Mg
Aluminium	Al
Zinc	Zn
Iron	Fe
Lead	Pb
Copper	Cu
Silver	Ag
Gold	Au

Most reactive → Increasing reactivity → Least reactive

Increasing ease of electron loss

Less easily corroded

Metals above this line react with acids, displacing hydrogen

recessive. The allele for brown eyes is dominant. A person will have blue eyes only when two blue-eye alleles make up the pair controlling eye colour. (see *heterozygous)

recrystallization A method of purifying some solid substances. The substance is dissolved in a *solvent and the solution that results is filtered and *crystallized. The crystals are removed from the solution and the process is repeated.

rectifier An electronic device that changes an alternating current into a direct current (*electric current). A *diode can be used as a rectifier.

rectum The part of the large *intestine between the *colon and the *anus or *cloaca. The rectum stores solid indigestible and other waste matter before it is removed from the body by defecation.

recycle To use old manufactured items as a source of *raw materials. Paper, glass, some plastics, and aluminium drinks cans are commonly recycled.

red giant A giant star in the last stage of its evolution, when its fuel is almost exhausted. Red giants are 10 to 100 times the size of the Sun.

redox reaction A *chemical reaction where electrons are transferred from one *reactant to another. This means that *oxidation and *reduction happen at the same time. E.g. the reaction between sodium and chlorine to produce sodium chloride. Chlorine (the *oxidizing agent) takes electrons from the sodium (the *reducing agent). As a result, chlorine is reduced and sodium is oxidized.

red shift An apparent change observed in the frequency of light that comes from distant stars. It is caused by the *Doppler effect that alters (shifts) the frequency of the light towards the red end of the *spectrum. Red shift in light from distant galaxies

255 **refinery**

shows that the Universe is expanding and is evidence for the *big-bang theory.

reducer see *reducing agent

reducing agent (reducer, reductant) A chemical *reagent that causes *reduction to happen to an atom or group of atoms by adding electrons to it. Reducing agents include hydrogen, carbon monoxide, and iron(II) (Fe^{2+}) ions.

reductant see *reducing agent

reduction A *chemical reaction where oxygen is removed from a substance. E.g. hydrogen reacting with copper oxide to produce copper and water. The copper in the copper oxide has gained electrons: $Cu^{2+}O^{2-}(s)+H_2(g) \rightarrow Cu(s)+H_2O(l)$. Reduction is therefore said to happen to any atom or group of atoms that gains electrons during a chemical reaction.

refinery see *oil refinery

Reflection

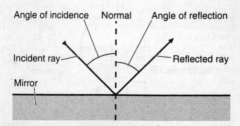

Angle of incidence   Normal   Angle of reflection

Incident ray

Reflected ray

Mirror

Reflux

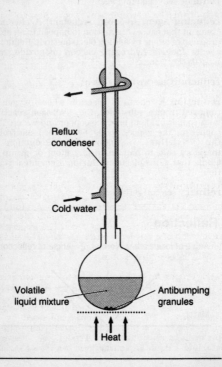

Reflux
condenser

Cold water

Volatile
liquid mixture

Antibumping
granules

Heat

refining Making a substance purer. Sugar is refined by *recrystallization. Impure copper is refined by *electrolysis.

reflection The return of *waves or a *beam of particles when they meet an obstacle. The angle of incidence equals the angle of reflection.   ✍

reflex An automatic movement or other response made by an animal. A human hand quickly pulls back if pricked by a pin. Information from a *receptor passes straight to an *effector without passing through the *brain.

reflux condenser A *condenser attached vertically to a container of boiling liquid. It allows volatile liquids (e.g. *ethanol) to boil for a long time without loss. Vapour rising into the condenser is changed back into a liquid and returns to the container.   ✍

refraction The change of direction of a *ray when it changes speed as it goes from one medium into another. The angle of refraction is less than the angle of incidence when the ray travels from a less to a more dense medium, and vice versa. The larger the wavelength of a wave, the larger the angle of refraction. Refraction makes water appear shallower than it really is. (see *dispersion)   ✍

refrigerator An appliance whose internal temperature is below that of its surroundings. A volatile liquid (usually a *chlorofluorocarbon CFC) evaporates and becomes a gas, removing heat from the freezer compartment. The gas is then compressed by a pump. The hot compressed gas flows through pipes outside the refrigerator where it cools and condenses back into a liquid. A refrigerator is a type of *heat pump.   ✍

Refraction

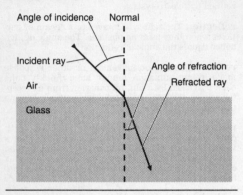

Angle of incidence   Normal

Incident ray

Angle of refraction

Refracted ray

Air

Glass

relative atomic mass (r.a.m.) The mass of an atom, measured on a scale where an atom of carbon-12 equals 12.0000. E.g. hydrogen equals 1; sodium equals 23. The r.a.m. of an element is the *average* mass of the atoms present in a naturally occurring sample of the element. E.g. The r.a.m. of chlorine is 35.5 because chlorine is a mixture of the *isotopes ^{35}Cl and ^{37}Cl.

relative molecular mass (r.m.m.) The *formula mass of a substance that is made up from molecules. It is calculated by adding the *relative atomic masses of the atoms in 1 molecule of the substance. E.g. ammonia (NH_3) r.m.m. is 14+1+1+1 = 17.

relay A switch that is controlled by an *electromagnet. A small current flows through the input circuit

Refrigerator

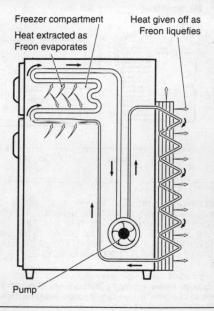

Freezer compartment

Heat extracted as
Freon evaporates

Heat given off as
Freon liquefies

Pump

and energizes an electromagnet which attracts an
iron armature. The armature pushes together two
contacts, allowing a large current to flow in the output circuit.

Relay

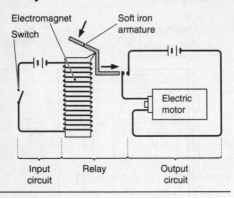

Electromagnet

Soft iron armature

Switch

Electric motor

Input circuit

Relay

Output circuit

renewable energy Energy that comes from sources that can be replaced naturally. Sources of renewable energy include *biofuels, *geothermal energy, *wave power, *tidal power, *wind power, *hydroelectricity, and *solar energy. Renewable energy is a form of *alternative energy.

reproduction The process used by living organisms to make copies of themselves. The two main types of reproduction are *asexual reproduction and *sexual reproduction.

reptile A type of *vertebrate animal. Reptiles include lizards, crocodiles, tortoises, and snakes. They have a dry and scaly skin, breathe with lungs, and lay eggs that have leathery shells.

resistance A property of an electrical *conductor to work against the flow of the current and change some of the electrical energy into heat. The *SI unit of resistance is the *ohm (Ω). Different substances have different natural values of resistance. For a given substance, the longer and thinner it is, the greater its resistance. Heating a conductor usually increases its resistance.

resistor A component included in some electric *circuits. Resistors have electrical *resistance and control the magnitude of the *electric current flowing. They are made from either a coil of thin wire or a *ceramic material containing carbon grains.

resonance The effect of a vibration with a small *amplitude making an object oscillate with a large amplitude. The resonant frequency is the frequency at which an object naturally vibrates. E.g. resonance makes the windows in a bus rattle when the engine is running at a particular speed.

resource Materials taken from the world around us and used as fuels or to make things. *Fossil fuels and metal *ores are non-renewable (finite) resources. They are steadily being used up and there are no new supplies. Renewable (non-finite) resources include wood and *renewable energy.

respiration External respiration moves oxygen from the space outside an organism to the cells inside its body (see *breathing). Carbon dioxide moves from the cells to the outside. Internal respiration (tissue respiration) takes place inside cells, releasing energy from *glucose. There are two forms of internal respiration: *aerobic respiration and *anaerobic respiration.

retina see *eye

reversible reaction A *chemical reaction where the *products react to re-form the *reactants. E.g.

$3H_2(g)+N_2(g) \Leftrightarrow 2NH_3(g)$. The forward reaction combines nitrogen and hydrogen to make ammonia: $3H_2(g)+N_2(g) \rightarrow 2NH_3(g)$. At the same time, the backward reaction *decomposes ammonia into hydrogen and nitrogen: $2NH_3(g) \rightarrow 3H_2(g)+N_2(g)$. (see *chemical equilibrium)

rheostat A type of variable *resistor that is used to control the size of the *electric current flowing in a *circuit.

rib One of the bones that make up the rib-cage that encloses the lungs and heart in vertebrates. (see *skeleton)

ribosome A type of *organelle found in a living *cell. Ribosomes manufacture *proteins from *amino acids. A type of *RNA from the *nucleus controls this process.

ripple tank A flat transparent tank used to show how *waves behave. A motor-driven 'dipper' sets up ripples (transverse waves) that travel across the tank. The ripples can undergo *diffraction, *reflection, and *refraction (just like light) when suitable barriers are used.

r.m.m. Abbreviation for *relative molecular mass.

RNA Ribonucleic acid, a *nucleic acid found in the *nucleus and *cytoplasm of a living cell. The production of RNA in the nucleus is controlled by *DNA in *chromosomes. RNA controls the manufacture of *proteins at *ribosomes.

rock A solid made from *mineral particles and found in the Earth's *crust. There are three main kinds of rock: *sedimentary, *igneous, and *metamorphic.

rock cycle The steady changing of one type of *rock into another. The changes take many millions of years to complete. ✍

Rock cycle

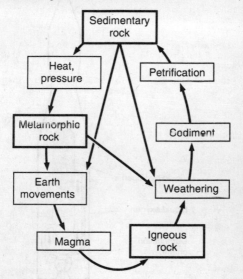

rocket A vehicle fitted with a *rocket engine, used to carry *artificial satellites and astronauts beyond the Earth's atmosphere into space. In a two-stage rocket, a large rocket lifts a smaller rocket above the atmosphere. The rockets separate, the empty larger rocket falls away, and the smaller one continues the journey.

Rocket engine

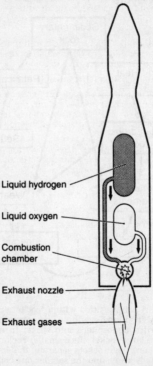

Liquid hydrogen

Liquid oxygen

Combustion chamber

Exhaust nozzle

Exhaust gases

rocket engine A type of *engine used to move space vehicles. It gives *jet propulsion by burning a mixture of fuel and *oxidizing agent in a combustion chamber. Rocket engines do not use oxygen from the air.   ✍

rod cell A type of light-sensitive cell in the retina of the *eye. Rod cells are responsible for vision in dim light. They are not sensitive to colour.

ROM Abbreviation for *read only memory.

root 1. The part of a *plant that grows downwards into the ground. The root anchors the plant and is used for food storage. Roots are covered in root hairs that take in water and dissolved *inorganic salts. 2. see *tooth.

rot Another term for *decompose, applied to the breaking down of dead plants and animals by *saprophytes and other *decomposers.

rotor The part of an *electric motor, *generator, or *alternator that rotates.

roughage see *dietary fibre

rusting The *corrosion of iron or steel by water and oxygen acting together. Rust is *hydrated iron oxide $Fe_2O_3.xH_2O$.

S

sac A bag-shaped part in a living organism. It is surrounded by a *membrane. A sac may contain a liquid or a gas or an *organ. The yolk in a hen's egg is contained in a sac.

saccharide see *sugar

sacrificial protection Protecting a metal from *corrosion by connecting it to a more reactive metal. E.g. magnesium connected to a steel pipeline and zinc attached to a ship's hull. The magnesium or zinc corrodes. Electrons flow from these metals into the steel and stop it corroding.

saliva A watery liquid made by glands in the mouth of land vertebrates. Saliva moistens food and makes it easier to chew and swallow. It also contains an *enzyme that starts the process of *digestion. (see *amylase)

salt A compound formed when an *acid and a *base react together, e.g.

$$H_2SO_4(aq)+CuO(s) \rightarrow CuSO_4(aq)+H_2O(l).$$
$$\text{acid} \qquad \text{base} \qquad \text{salt} \qquad \text{water}$$

Salts are solids made up from ions. Examples include sodium chloride NaCl, copper carbonate $CuCO_3$, and ammonium nitrate NH_4NO_3. (see *radical)

sampling In *ecology, selecting objects or organisms within a small area and counting or measuring them. Sampling is usually done at several random places across a *habitat. The results are used to calculate information about the whole of the habitat. (see *quadrat)

sand Particles of *rock between 0.05 and 2 mm in diameter. Grains of sand are mostly silicon dioxide and come from the *weathering of rocks made of *quartz.

sandstone A *sedimentary rock made of sand.

sap The liquid in a *plant. Sap is water containing dissolved substances. It flows through pipes called *xylem and *phloem vessels.

saprophyte An organism that feeds on other dead organisms. Saprophytes release *enzymes that decompose their food into a solution which they can absorb. They are usually *fungi or *bacteria and are *decomposers that play an important part in the *nitrogen and *carbon cycles. Saprophytes release *nutrients that can be used by growing plants.

satellite An object that moves in an *orbit around a planet. E.g. the Earth has one natural satellite, the Moon. (see *artificial satellite)

saturated 1. Describing a *solution that can dissolve no more *solute, at that temperature. 2. Describing an *organic compound that contains only single bonds. Saturated compounds take part in *substitution reactions. All *alkanes are saturated.

scalar A measurement that shows a quantity or amount but that does not show a direction. Mass, speed, and density are scalar quantities. (see *vector)

scale 1. A series of lines and numbers on a measuring instrument. An indicator moves across the scale to show the reading to be taken. *Analogue thermometers, voltmeters and ammeters, and burettes and measuring cylinders have scales. 2. How much of an actual quantity is represented by a certain length or area on a diagram. E.g. on a 1:100 scale diagram, a pipe 1 m in diameter is drawn 1 cm wide; on a *bar chart showing student height in a school, a 1 cm square may represent 5 students.

scapula The bone that makes up the shoulder-blade of humans and many other vertebrates. (see *skeleton)

sclerotic layer see *eye

screw Part of a simple *machine, consisting of a continuous groove (thread) around the outside of a cylinder or cone. The thread is effectively a spiral *inclined plane. The groove (thread) on the outside of a bolt engages with the thread inside a nut. Rotat-

Screw

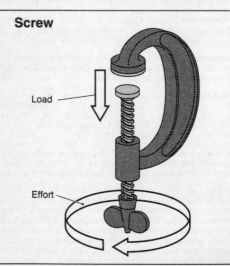

Load

Effort

ing a bolt moves the nut along it with great force. A jack uses a screw to lift a heavy car easily.   ✍

scrotum In male mammals, the sac of skin that contains the two *testes (testicles). The scrotum lies outside the body and keeps the testes at the lower temperature needed for the formation of *sperms. (see *sexual reproduction)

scrubber A tall tower packed with glass balls or bricks. Scrubbers are used in industry to remove impurities from gases. The gases pass up the tower as water trickles down and removes the impurities.

sebaceous gland see *skin

sebum see *skin

second (symbol s) The *SI unit of time. There are 60 s in 1 minute and 3600 s in 1 hour.

secondary cell An electrochemical *cell (e.g. *nicad and *lead–acid cells) that can be recharged when exhausted (flat). A secondary cell is recharged by passing an electric current in the opposite direction to the current that flows when the cell is in use. The charging current reverses the *chemical reactions that happened during use and restores the contents of the cell to their original states.

sediment Solid particles, originally in a *suspension, that settle downwards and form a layer. Layers of sediment build up on ocean floors as rivers carry weathered rock into the sea. The weight of the layers compresses the sediment in them and makes *sedimentary rocks.

sedimentary rock A type of *rock, formed when separate *particles of other rocks settle in layers and become compacted and stuck together. Sedimentary rocks are usually soft and wear away easily. Examples include *sandstone and *chalk. (see *deposition *weathering)

sedimentation The process of small particles of solid in a *suspension settling to the bottom of the liquid.

seed A seed grows in a *plant when male *pollen fertilizes a female *ovule. Seeds have an outer covering called the testa. The testa contains an *embryo made up from a *plumule, a *radicle, and one or two *cotyledons. During *germination, the embryo grows into a new plant. Some seeds also contain endosperm, which is a store of starchy food.

seismic wave A *wave that travels through the Earth (at between 4 and 8 km/s) away from sudden underground rock movements. The ground shakes and shudders when seismic waves reach the surface and the result is an earthquake.

seismometer A device used to indicate earthquake waves. It is made of a suspended mass with high *inertia that stays still while the ground it is supported on moves.

selection The effect of the *environment on an organism, causing it to live and be successful or to die. It depends on the factors in the environment and the *characteristics of the organism.

semen A liquid made by male animals that contains *sperms. It is used during *sexual reproduction to help carry the sperms to the female *ova.

semicircular canal see *ear

semiconductor A substance whose electrical *conductivity is midway between the conductivity of a metal (e.g. copper) and an insulator (e.g. glass). Crystals of *silicon and germanium are specially treated to make the semiconductor material used in *transistors, *diodes, and *integrated circuits.

seminal vesicle A *gland connected to the *vas deferens in male mammals. It produces a liquid that mixes with *sperms to make *semen. (see *sexual reproduction)

semi-permeable membrane A *membrane that contains microscopic holes or pores. During *osmosis, small *solvent molecules pass freely through the pores. Larger *solute molecules or ions cannot pass. Examples of semi-permeable membranes are specially treated porous pots, Visking tubes and *cell membranes.

sense What an organism uses to take in information about the world around it. The human senses of sight, hearing, taste, and smell depend on *sense organs. The sense of touch depends on *receptors in the *skin that are sensitive to pressure, touch, pain, and temperature.

sense organ In humans, the *eyes, *ears, *tongue, and *nose. Sense organs contain *receptors that send information to the *brain.

sepal see *calyx

series connection A method of connecting together separate components in an electric *circuit so that the *electric current passes through each component in turn. Two similar bulbs connected in series have the battery *voltage divided equally between them. The current is half the value it would be for a single bulb. Two 6-*volt batteries connected in series have a voltage of 12 volts but can supply only the same current as one 6-volt battery. (see *parallel connection)   ✍

serum Blood *plasma from which substances that cause blood to *clot have been removed. Serum containing particular *antibodies can be injected into people to treat or prevent some diseases. (see *immunize)

sex cell see *gamete

Series connection

Bulbs in series

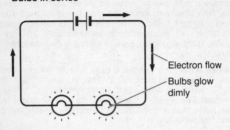

Electron flow

Bulbs glow dimly

Cells in series

6V

2V 2V 2V

sex chromosome A *chromosome that decides the sex of an organism. There are two sex chromosomes in mammals, X and Y. Females have two similar X chromosomes. Males have an X chromosome and a Y chromosome. A female *ovum has an X chromosome. A male *sperm has either an X or a Y chromosome. The sex of offspring is decided by the type of sperm that fertilizes the ovum.

sex linkage Inherited *characteristics carried on the X *sex chromosome and occurring more often in one sex than another. E.g. *colour-blindness affects more men than women. (see *haemophilia)

Sexual reproduction

Male

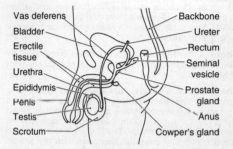

Vas deferens
Bladder
Erectile tissue
Urethra
Epididymis
Penis
Testis
Scrotum

Backbone
Ureter
Rectum
Seminal vesicle
Prostate gland
Anus
Cowper's gland

Female

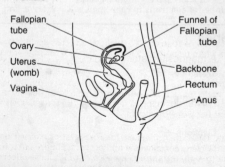

Fallopian tube
Ovary
Uterus (womb)
Vagina

Funnel of Fallopian tube
Backbone
Rectum
Anus

sexual reproduction A type of reproduction where a *gamete from a female organism joins with a gamete from a male organism. This forms a *zygote which grows into a new organism. Male and female organisms have special sexual (reproductive) organs which produce the gametes. In humans and other mammals, the female's *ovaries produce gametes called *ova and the male's *testes produce *sperms. (see *fertilization *flower) ✍

shale A *sedimentary rock formed from mud or clay. Slate is the *metamorphic rock that forms from shale.

shell (electron shell) The space that contains a group of *electrons orbiting around the *nucleus of an *atom. All the electrons in a shell orbit at about the same distance from the nucleus and have approximately the same energy. The K shell is the closest to the nucleus and contains electrons with the lowest energy. The K, L, M, etc. shells are at successively greater distances from the nucleus and contain electrons of successively higher energy. The maximum number of electrons in each shell is K = 2, L = 8, and

Shell

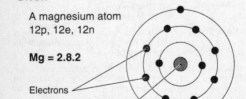

A magnesium atom
12p, 12e, 12n

Mg = 2.8.2

Electrons

Nucleus containing
12 neutrons and 12 protons

Short circuit

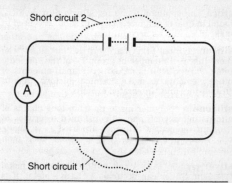

Short circuit 2

A

Short circuit 1

M = 8. The outer shell of one atom can overlap with the outer shell of another atom to form a *covalent bond. One or more pairs of electrons are shared between the two atoms. ✍

shoot Part of a plant above the ground made up from a *stem supporting leaves, buds, and flowers. It develops from the *plumule in a seed.

short circuit A fault in an *electric circuit. A short circuit bypasses part of the circuit and can cause an unusually large current to flow. ✍

short sight A person with short sight cannot see distant objects clearly. The *eye focuses them in front of the retina. Short sight is overcome by wearing spectacles with *concave lenses.

sickle cell anaemia A genetic (inherited) disease which causes red *blood cells to contain defective

silica

*haemoglobin. The disease is carried by a *recessive allele.

silica The common name for silicon dioxide SiO_2, found in quartz and many types of sand. It is used to make glass and furnace linings.

silicon (symbol Si) A hard brittle *metalloid *element that is a member of group IV of the *periodic table. Pure silicon is mixed with small amounts of other elements to make *semiconductors used in *transistors and *integrated circuits.

silicone A *polymer made up from long chains of alternating oxygen and silicon joined to groups of *organic molecules. Silicones are used as synthetic oils, waxes, and rubbers. They are more *inert than the natural substances.

silver (symbol Ag) A soft white *transition metal *element. Silver can occur as a *native metal. It is usually extracted as a *by-product during the *smelting and *refining of copper and lead. It is used to make jewellery and photographic film.

sintering Heating and compressing a powdered solid at a temperature below its melting-point. The particles of solid stick together to form a larger porous lump. Sintering can be used with glass, powdered *ores, *alloys, and *ceramics.

sinus A space inside the bone or other *tissue of an animal. Humans have air-filled sinuses which are inside the bones of the skull. They connect to the nasal cavity. The two largest sinuses are in the bones of the forehead.

SI units The standard set of internationally agreed units of measurement. The three basic units of length, mass, and time are the *metre, the *kilogram, and the *second. Most other units are based on these.

Skeleton

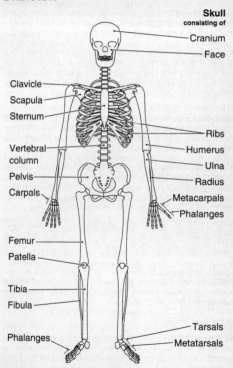

Skull
consisting of
- Cranium
- Face

Clavicle
Scapula
Sternum

Ribs
Humerus
Ulna
Radius

Vertebral column

Pelvis
Carpals

Metacarpals
Phalanges

Femur
Patella

Tibia
Fibula

Tarsals
Metatarsals

Phalanges

skeleton

skeleton The hard part of an animal that supports
the body and protects internal *organs. *Muscles are
attached across *joints in the skeleton and allow
parts of the skeleton to move. Some *invertebrates
have an external skeleton (exoskeleton) that totally
encloses the body and limbs. *Vertebrates have an
internal skeleton (endoskeleton) made of *bones or
*cartilage. The axial skeleton is made up from the
bones of the head and body. The appendicular skeleton contains the limbs. Some animals have a skeleton made up from *tissues containing liquid under
pressure, e.g. earthworms, slugs, caterpillars. ✍

skin The outer covering of a *vertebrate. It consists
of two parts, the epidermis and the dermis. The epidermis of a mammal has an outer layer of dead cells
which prevents water loss and the entry of bacteria.
These cells steadily wear away and are replaced
from a lower layer of living cells. Sebaceous glands
in the dermis secrete oily sebum which lubricates
and waterproofs the skin and hair. The dermis also
contains nerve endings that sense temperature, pressure, pain, and touch. Blood *capillaries and hairs
assist *homoeostasis by regulating heat loss. Sweat
glands excrete sweat, which is water containing salts
and *urea. Evaporation of sweat cools the skin and
cools blood in the capillaries. ✍

skull The bones that make up the head of a *vertebrate animal. In humans, the skull contains the *cranium and the bones of the face and jaw. (see
*skeleton)

slag A substance containing impurities and made
during the *smelting of metals. Slag from a *blast-
furnace is calcium silicate. It forms when silicon
dioxide in iron *ore reacts with calcium oxide from
*limestone.

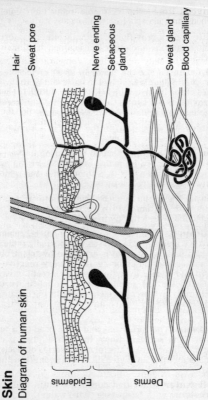

Skin
Diagram of human skin

Hair · Sweat pore · Nerve ending · Sebaceous gland · Sweat gland · Blood capilliary

Epidermis · Dermis

slaked lime The common name for calcium hydroxide, $Ca(OH)_2$. (see *lime)

slate A dark grey flaky *metamorphic rock formed from mudstone and other fine-grained *sedimentary rocks.

slip ring Part of the *rotor in an electric *alternator. A slip ring is attached to each end of the coil of wire that spins with the rotor. Fixed carbon *brushes press against each slip ring and conduct electrical energy away from the spinning coil.

slurry A *suspension in the form of a semi-liquid paste that is able to flow.

smelting Obtaining a metal by heating its *ore in a furnace. The ore is heated with a *reducing agent (e.g. *coke), and a flux (e.g. *limestone) that removes impurities. (see *blast-furnace)

soap A *detergent made by heating natural *fats and *oils with sodium hydroxide. Soaps are *fatty acid molecules with long *hydrocarbon chains.

sodium (symbol Na) A soft silvery *metal *element that is a member of group I in the *periodic table (the *alkali metals). Sodium is extracted by *electrolysis of molten sodium chloride. It is a very reactive element and is used in sodium vapour street lamps and as a coolant in some types of *nuclear reactor. Sodium is an *essential element for living organisms and many of its *salts have important industrial uses.

software The *programs and *data used by a *computer. The software flows through a *digital computer's *hardware in the form of a series of electrical pulses making up *bytes of information. (see *hardware)

soft water Water that does not contain *salts that form scum with *soap. (see *hard water)

Soil

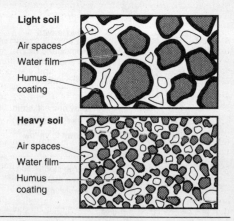

Light soil

Air spaces

Water film

Humus
coating

Heavy soil

Air spaces

Water film

Humus
coating

soil A mixture of rock particles, *humus, air, and
water that provides a suitable environment for growing plants. (see *loam)

soil depletion Loss of plant *nutrients from soil.
Soil depletion is caused by *leaching and by regularly growing crops without the use of *crop rotation
or *fertilizers.

soil erosion The removal of soil by wind or water.
Growing plants protect the surface of the soil and
their roots help to bind it together. Soil erosion happens when soil is exposed by deforestation or by
repeated ploughing and harvesting of crops.

Soil profile

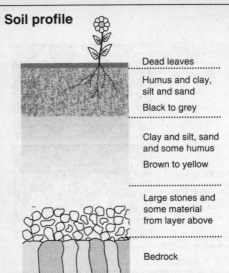

Dead leaves

Humus and clay,
silt and sand

Black to grey

Clay and silt, sand
and some humus

Brown to yellow

Large stones and
some material
from layer above

Bedrock

soil profile Layers of different kinds of soil. The soil profile can be seen by digging a hole with straight sides.

solar cell A type of *photocell that generates an *electric current when sunlight falls on it. Solar cells are made from wafers of silicon. A cell 10 cm in diameter can produce about 0.75 *watts of power in bright sunlight. Solar panels made up from solar cells connected together are used to power equipment in *artificial satellites. Solar panels are also used, especially in isolated places, on Earth as sources of *alternative energy.

solar energy Energy from the Sun. Solar energy is the original source of all forms of energy on Earth except *nuclear energy. Solar energy powers *photosynthesis, the *water cycle, and changes in the weather. It was responsible for the formation of *fossil fuels.

solar panel see *solar cell

solar radiation The particles and *waves given off by the Sun. Solar radiation is made up from the *solar wind together with the whole *spectrum of *electromagnetic radiation.

solar system The Sun and everything that moves around it. The solar system includes the nine planets and their *moons, as well as *asteroids and *comets.

solar wind A stream of particles (mostly protons and electrons) that continuously escape into space from the Sun. The solar wind affects the Earth's *ionosphere and so influences the reception of radio waves.

solder An *alloy used to join other metals together. Soft solder contains *tin and *lead, melts at about 250 °C, and is used to join wires and components in electronic circuits. Brazing solder contains tin and *copper, melts at about 800 °C, and is used to join pieces of steel.

solenoid A long thin coil of wire. A *magnetic field surrounds the coil when an electric current is passed through it. The field has the same shape as that surrounding a bar *magnet. Pushing an iron core into a solenoid increases the intensity (strength) of the field. (see *electromagnet)

Solar system

Name	Distance from sun (x 10⁶ km)	Period of orbit	Diameter (x 10³ km)	Mass with respect to Earth (Earth = 1)
Mercury	58.0	88 days	5.0	0.0543
Venus	108.2	224.75 days	12.4	0.8136
Earth	149.7	365.25 days	12.762	1.000
Mars	227.8	687 days	6.8	0.108
Jupiter	778.1	11.86 years	142.8	318.35
Saturn	1426.6	29.46 years	120.9	95.3
Uranus	2871	84.01 years	47.2	14.58
Neptune	4497	164.79 years	44.6	17.26
Pluto	5893	248.43 years	5.8	0.7

solid A *state of matter where a substance is rigid and keeps its shape. The *particles in a solid are closely packed together in fixed positions. (see *lattice)

solubility The maximum amount of a *solute that will dissolve in a fixed volume of *solvent at a stated temperature. Units are usually grams of solute per dm^3 of solution.

soluble Describing a substance that is able to dissolve in a liquid to form a *solution.

solute The substance that dissolves in a *solvent (liquid) to form a *solution. E.g. salt is the main solute in sea water; carbon dioxide is a solute in lemonade.

solution A uniform mixture that is in one state and consists of a *solute dissolved in a *solvent. A solution in water is clear; you can see through it.

solvent The liquid that dissolves a *solute to form a *solution. E.g. water is the solvent in salt solution.

sonar Equipment used to measure the depth of water under a ship and to locate submarines and shoals of fish. It builds up a picture by sending out pulses of sound in different directions and measuring the time for each *echo to return.

sound *Energy that moves as longitudinal *waves through a medium (e.g. air) and that can be detected by the ears of an animal. Sound moves through air at a *velocity of 344 metres/second (m/s) and through water at 1461 m/s. Young people can hear sounds with *frequencies between 30 and 20000 hertz (Hz). (see *ultrasound *infrasound)

space probe An unmanned spacecraft that gathers information from planets or moons or other bodies in the *solar system. Radio signals carry the information back to Earth.

species A group of living organisms that are able to breed with each other and produce *fertile offspring. Several different species make up a *genus. Lions are one species and tigers are another species. Horses and donkeys belong to different species. (see *classification)

spectrum A range of related things, listed in a set order. White *light can be split into a spectrum of different colours. (see *electromagnetic spectrum)

speed A measure of how fast an object is moving. Speed tells you how far something moves in a unit of time.

$$\text{Speed} = \frac{\text{distance moved}}{\text{time taken}}.$$

Units of speed are metres per second (m/s) and kilometres per hour (km/h).

sperm (spermatozoon) The *gamete made in the *testes of male animals. Most sperms are shaped like a tadpole and move with the help of a *flagellum. Sperms join with *ova during *fertilization.

spermatozoon see *sperm

sphincter A ring of muscles around a tube inside the body of an animal. The sphincter muscles contract and squeeze the pipe to stop the flow of substances through it. The *stomach, *anus, and urinary *bladder have sphincters at their outlets.

spinal column see *vertebral column

spinal cord In *vertebrates, the bundle of *nerves that runs from the *brain and down through the inside of the *vertebral column (spine). The spinal cord carries nerve impulses between various parts of the body and the brain.

spine 1. A sharp protective point that grows on the leaves of some *plants (e.g. holly). 2. see *vertebral column

spiracle A small hole on the side of an *insect or other *arthropod. Air enters and leaves the animal's body through rows of spiracles.

spleen In vertebrates, an organ connected to the blood *circulation. It lies close to the stomach. The spleen makes *lymphocytes, stores red *blood cells, and removes worn red blood cells from the circulation.

spore A simple type of cell used by *fungi and some simple plants for reproduction. Spores are produced by *vegetative reproduction and grow into new organisms without the need for *fertilization.

stability 1. A measure of how well something can resist attempts to change its position. A wide structure with a low *centre of gravity has greater stability than a narrow structure with a high centre of gravity. 2. A measure of how well a substance can resist attempts to make it take part in a *chemical change. E.g. calcium carbonate has greater stability to heat than copper carbonate because it *decomposes at a higher temperature.

stamen see *flower

standard temperature and pressure (s.t.p.) Fixed conditions used when comparing the properties of gases. Standard temperature is 273 K (0 °C). Standard pressure is 101 325 Pa (760 mm of mercury).

stapes see *ear

star In astronomy, a large object that gives out energy from *nuclear fusion. The Sun is a small star that produces its energy from the fusion of hydrogen into helium.

State

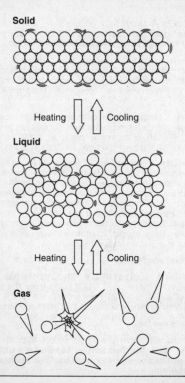

Solid

Heating   Cooling

Liquid

Heating   Cooling

Gas

starch A *carbohydrate stored as food by plants and used by animals to produce energy. Pure starch is a white powder that does not dissolve in water. Each molecule of starch is made up from thousands of *glucose molecules joined in a long chain.

state The form in which a substance exists, either a *solid, a *liquid, or a *gas. (see also *plasma)   ✍

static electricity The effects caused by charged particles (electric *charge) collecting in one place. Static electricity makes hair stand on end when combed in dry weather. A stroke of lightning is a sudden flow of what was static electricity between a charged cloud and the ground.

stationary wave The result of *interference (between two *progressive waves) when a *wave is reflected back on its own path. A stationary wave has fixed points where the *amplitude is zero (compared to a progressive wave where the points of zero amplitude move at the speed of *propagation of the wave). Stationary waves are set up in musical instruments, e.g. in vibrating strings (piano, guitar); in columns of air (trumpet, recorder).

stator In an *electric motor, *generator, or *alternator, the non-moving part that provides the magnetic *field in which the *rotor spins.

steam engine An *engine that uses high-pressure steam to force a *piston along a cylinder. The piston is connected to a *crankshaft that turns a *flywheel. The steam comes from water heated in a boiler by a furnace burning a fuel.

steam reforming An industrial process, using a nickel *catalyst to react *methane with steam to obtain *hydrogen and *carbon monoxide.

steel An *alloy containing *iron mixed with other *elements. Mild steel is made by blasting oxygen into molten *pig iron and contains up to 0.5% carbon. It is used to make girders, ships, and vehicle bodies. Other metals are added to iron to change its properties. Adding *chromium and *nickel makes unreactive stainless steel. Adding manganese makes tough and springy steel. Adding *tungsten makes hard steel used in cutting tools.

stem In *plants that grow from seeds, a part that supports leaves and *buds. Most types of stem grow above the ground.

sterile 1. A living organism is sterile if it cannot reproduce. 2. An object is sterile if it is free from living *micro-organisms. Doctors and surgeons use sterile equipment during operations.

sternum The breastbone that joins the *ribs at the front of the *thorax (chest) in many vertebrates. (see *skeleton)

stigma see *flower

stimulus (pl. stimuli) Anything that affects the inside or the outside of an organism and which triggers it to make a response. Stimuli are received by *organs called receptors. Animals such as humans have *sense organs which are the receptors for external stimuli such as light, sound, or touch. These sense organs send messages along nerves to the brain. The effects of many internal stimuli (e.g. hormones) are often not consciously noticed.

Stimulus	Organism	Response
sound of crying baby (*external*)	mother	feeds baby
food in stomach (*internal*)	baby	glands secrete digestive juices
growth hormone in blood (*internal*)	child	bones grow longer
bright light (*external*)	plant	grows towards light

stoma (pl. stomata) A small hole in the skin of a *leaf. Carbon dioxide passes into the leaf through the stomata. Water vapour and oxygen pass out. The size of stomata is controlled by *guard cells.

stomach An *organ in vertebrates that helps to digest food. The stomach is a muscular bag that squeezes swallowed food and mixes it with digestive juices. These juices are made by glands in the stomach wall and contain hydrochloric acid and pepsin. (see *alimentary canal)

s.t.p. Abbreviation for *standard temperature and pressure.

strain A group of organisms from the same *species but with different *characteristics. Different strains (breeds) of rabbit can have large or small ears. Different strains (varieties) of rose can have red or white petals.

strata (sing. stratum) Layers (bands) of *sedimentary rocks. Strata are horizontal when sedimentary rocks are first formed. (see *fold *fault *deposition)

streamlined Describing the shape of an object that has low *drag. Submarines, sharks, and most modern cars have streamlined shapes.

strength 1. The ability of a material to resist the effect of a force. E.g. *tensile strength* is a measure of the force required to break an object by pulling at opposite ends. 2. A measure of the magnitude of a force or an effect. E.g. the strength (*pressure*) of a jet of water or the strength (*amperage*) of an electric current. 3. The *concentration* of *solute in a *solution. (NB the word 'strength' is best avoided in science; where possible, use the *more precise* terms, as shown.)

structure The overall shape of something and the arrangement of its parts. The structure of an *atom is described by the arrangement of its *electrons and *nucleons. The structure of a molecule is described by its geometrical *formula and of a *crystal by the arrangement of its *ions or molecules in a *lattice. The structure of living cells includes their *organelles.

style see *flower

sublime To undergo a *physical change where a solid changes straight into a gas when heated, without first melting to a liquid. Iodine and ammonium chloride sublime.

substance A sample of *matter that can be identified by name, e.g. water, cotton, crude oil, methane. Substances exist in three main states: *solid, *liquid, and *gas.

substitution reaction A *chemical reaction where one or more atoms in a molecule are replaced by other atoms. E.g. the reaction between *ethane and *bromine:

$$
\begin{array}{c}
\quad H \quad H \qquad\qquad\qquad H \quad H \\
\quad | \quad\ | \qquad\qquad\qquad | \quad\ | \\
H - C - C - H + Br_2 \rightarrow H - C - C - Br + HBr. \\
\quad | \quad\ | \qquad\qquad\qquad | \quad\ | \\
\quad H \quad H \qquad\qquad\qquad H \quad H
\end{array}
$$

succession A series of *communities that live in an area over time. Each community changes the *environment and makes it more suitable for the following community.

sucrose (formula $C_{12}H_{22}O_{11}$) A *sugar that is present in many plants (e.g. sugar-cane, sugar-beet). One *molecule of sucrose is made up from one molecule of *glucose joined to one molecule of another sugar called fructose.

sugar 1. A type of solid crystalline *carbohydrate that dissolves easily in water and has a sweet taste.

Sugars include *sucrose from sugar-cane and sugar-beet, fructose and *glucose from fruit, and lactose from milk. 2. The common name for sucrose.

sulphate An *ion and an acid *radical with the formula SO_4^{2-}. Sulphates form salts, e.g. zinc sulphate $ZnSO_4$. Most sulphate salts are soluble in water.

sulphide An *ion and an acid *radical with the formula S^{2-}. Sulphides form salts, e.g. iron sulphide FeS. Sulphide salts give *hydrogen sulphide when heated with a strong *acid.

sulphite An *ion and an acid *radical with the formula SO_3^{2-}. Sulphites form salts, usually containing *alkali metals, e.g. sodium sulphite Na_2SO_3. Sulphite salts are soluble in water and give *sulphur dioxide when heated with a strong *acid.

sulphur (symbol S) A brittle yellow *non-metal *element that is a member of group VI of the *periodic table. It occurs *native and in *sulphide *ores. Sulphur burns in air to form *sulphur dioxide and is used to make *sulphuric acid. There are two *allotropes of sulphur: rhombic and monoclinic sulphur. Both contain ring-shaped S_8 molecules. Sulphur compounds as impurities in coal are a major cause of *acid rain.

sulphur dioxide A choking acrid colourless gas with the formula SO_2, made by burning sulphur. It results from the *combustion of *fossil fuels that contain sulphur impurities and causes *acid rain.

sulphuric acid A strong *acid with the formula H_2SO_4, made by the *contact process. It is used in the manufacture of *fertilizers, paints, fibres, and detergents.

Sun The star at the centre of our *solar system. The Sun's diameter is 110 times larger than the Earth's, and its mass is 300 000 times greater. The Sun gives off *solar radiation that results from *nuclear fusion.

superconductor A conductor that has no electrical *resistance when cooled to a very low *temperature (usually below 23 K). No electrical energy is lost as heat when an *electric current flows in a superconductor. Small but extremely powerful *electromagnets and experimental *electric motors and *generators use superconductors.

surface tension An effect in the surface of a liquid that makes it behave as if it had a skin. Surface tension results from forces of attraction between the molecules in the liquid. It causes the spherical shape of liquid drops and allows some insects to walk on water.

suspension A mixture made up from small *insoluble solid particles shaken with a liquid. The particles stop light passing easily through the liquid. When a suspension is allowed to stand undisturbed, the particles settle downwards. A suspension is not clear; you cannot see through it.

suspensory ligament see *eye

sweat gland see *skin

switch A device that can interrupt the current flow in a *circuit. A switch has a part which moves to make a gap in a conductor that is part of the circuit.

symbiosis When two organisms from different species benefit from living with each other. The guts of cows contain *bacteria that help to digest the *cellulose in their food. The cows provide a safe environment for the bacteria. Cows and bacteria live in symbiosis.

synapse The space between the *dendrites where the ends of different *neurones meet. Nerve impulses

Synapse

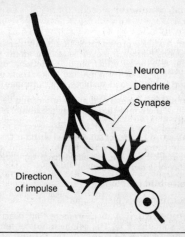

Neuron
Dendrite
Synapse

Direction
of impulse

---

pass from one neurone to another across the synapse.

syncline see *fold

synovial capsule A *membrane between two bones that move in a *joint. A synovial capsule is shaped like a bag. It contains a slippery liquid called synovial fluid that lubricates the joint.

synthesis Making chemical *compounds from *elements or more simple compounds.

systole see *blood pressure

T

tar A sticky brown-black liquid mixture of *hydro-carbons containing over 50 carbon atoms per mole-cule. It is the residue from the *fractional distillation of *crude oil or from making *coke from coal. Tar is used to make road surfaces. (see *bitumen)

tarsal In humans and many other land vertebrates, one of the bones that make up the ankle. (see *skele-ton)

tar sand Layers of sand mixed with tar that lie at or near the surface of the Earth. A liquid similar to *crude oil can be extracted from tar sand but the process is not economical.

taste bud see *tongue

tectonics see *plate tectonics

temperature A measurement that describes how hot something is. The common unit of temperature is the *degree Celsius (°C). The *SI unit of tempera-ture is the kelvin (K). (see *absolute zero *heat)

tendon A tough piece of *tissue that attaches a vol-untary *muscle to a bone. The muscle has at least two tendons, one at each end. In humans, the Achilles tendon joins the heel to the calf muscle.

tension A force that pulls. A weight hanging at the end of a rope causes tension in the rope. Bending a bar causes tension in the outside of the bend and *compression on the inside of the bend.

terminal velocity The maximum velocity reached by something that is falling. At the terminal velocity, the upward force of *drag equals the downward *gravitational force.

testa The outside covering of a *seed.

testicle The *testis of a male mammal. A mammal has two testicles, which are usually contained outside the body in the *scrotum. (see *sexual reproduction)

testis (pl. testes) An organ that produces *sperms in male animals. In mammals, it also produces the *hormone *testosterone.

testosterone A *hormone made in the *testes of male mammals. It gives adult males their secondary sexual characteristics. In humans, testosterone causes adult males to have deeper voices than females and more hair on their faces and bodies.

tetraethyl lead A poisonous liquid added to petrol to make it burn more efficiently in engines. Particles containing *lead escape in the engine exhaust fumes and cause *pollution.

theory An idea or a set of ideas used to explain the causes of connected observations. E.g. the *kinetic theory of matter, *Darwinism.

thermal decomposition A type of *chemical reaction where heat causes a *compound to *decompose. E.g. the thermal decomposition of sugar gives carbon and water.

thermal energy *Energy possessed by a body due to the motion of its atoms. A body increases its temperature or changes *state when it takes in thermal energy. (see *heat *absolute zero)

thermal expansion The increase in size of something as its temperature is increased. Thermal expansion is caused by the increase in speed of the atoms in a substance caused by their increase in

Thermal expansion

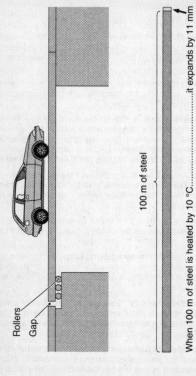

Rollers

Gap

100 m of steel

When 100 m of steel is heated by 10 °C............it expands by 11 mm

energy at the higher temperature. Sliding joints that allow for expansion in hot weather are included in railway lines and long bridges. Most *thermometers use the thermal expansion of a liquid or solid to indicate temperature rises. ✍

thermionic emission The release of electrons by some substances when heated. Thermionic emission is used in *thermionic valves and in *cathode-ray tubes.

thermionic valve An electronic device that can be used as a switch or an *amplifier. A triode valve uses a negatively charged wire grid to control the flow of electrons between a heated *cathode and a positively charged *anode. *Transistors have now mostly replaced thermionic valves in all electronic appliances except high-power radio and TV transmitters. (see *thermionic emission)

thermistor A type of *resistor whose resistance falls as its temperature rises.

Thermit reaction The highly *exothermic reaction between powdered iron oxide and aluminium powder: $Fe_2O_3 + 2Al \rightarrow Al_2O_3 + 2Fe.$

The molten iron that results can be used to weld steel objects, e.g. railway lines.

thermocouple A device that generates a small *electric current when one part is hotter than another part. A thermocouple is made up from two different metals joined together. A thermocouple connected to an *ammeter can be used as a thermometer to measure very high temperatures. ✍

thermometer A device for measuring *temperature. Glass thermometers contain liquid mercury or alcohol that expands as the temperature rises, and contracts as the temperature falls. The liquid moves along a narrow tube marked with a scale.

Thermocouple

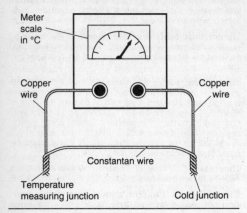

Meter scale in °C

Copper wire

Copper wire

Constantan wire

Temperature measuring junction

Cold junction

thermonuclear fusion see *nuclear fusion

thermoplastic see *plastic

thermoset see *plastic

thermostat A device that keeps a place at a steady temperature by controlling heating or cooling. Thermostats containing *bimetal strips control ovens, room heaters, and refrigerators. A thermostat switches an electric room heater off when the room is warm enough. When the temperature falls below a set value, the thermostat switches the heater on again.

thiosulphate An *ion and an acid *radical with the formula $S_2O_3^{2-}$. Thiosulphates form salts with *alkali metals, e.g. sodium thiosulphate $Na_2S_2O_3$. Thiosulphate salts give a *precipitate of sulphur when mixed with a strong *acid. Sodium thiosulphate is used in processing photographic film and prints.

thorax In vertebrates except fish, the part of the body that contains the heart and the lungs. In mammals, the thorax is enclosed by the rib-cage and is separated from the *abdomen by the *diaphragm. In *insects, the thorax bears the wings and legs.

thrust The forward *force that results from the stream of air or gases ejected backwards by a propellor, *turbojet, or *rocket engine. (see *reaction *jet propulsion)

thyristor A type of *semiconductor *rectifier which uses a small control current to switch a large current on and off. Variable-speed electric drills and lamp dimmers contain thyristors.

thyroid gland An *endocrine gland in vertebrates, found in the lower part of the neck. It makes a *hormone called thyroxine which controls the speed of chemical reactions (metabolic rate) in cells.

thyroxine A *hormone. (see *thyroid gland)

tibia The shin-bone, the larger of the two bones that join the ankle to the knee in humans and many other vertebrates. (see *skeleton)

tidal power Electrical energy generated from the energy of tides flowing in the sea. A barrier is built across an estuary. The tides flow in and out of the estuary through channels in the barrier, turning turbines that drive electric *generators. (see *renewable energy)

tidal volume The volume of air breathed in and out by a person at rest. The tidal volume of a teenager is about 400 cm³.

tide The regular rise and fall in the level of the Earth's oceans. The interval between high tide and the next low tide is approximately six hours. Tides are caused by the gravitational pull of the Moon and the Sun. When the Moon is either full or new, the Sun, Moon, and Earth are in a straight line. This causes spring tides, when there is a large difference in water-level between high and low tides. When the Moon is half-full, the Sun, Moon, and Earth are at right angles. This causes neap tides, when there is a small difference between high and low tides.

tin (symbol Sn) A very soft silvery *metal *element that is a member of group IV in the *periodic table. It is extracted from its *ore by *smelting. Tin is used in alloys (e.g. *solder) and to coat the thin steel plate in food cans. The coating stops *corrosion of the steel by the food.

tinplate A flexible sheet of steel *electroplated with a thin protective coat of tin. It is used to make food cans.

tissue A group of similar living cells that carry out the same job or function. Muscle tissue contracts and nerve tissue relays impulses (messages). An *organ is made up from different tissues that work together.

tissue fluid Also called interstitial fluid, a liquid that comes from blood. As blood circulates, tissue fluid leaks through the walls of the *capillaries. It carries food and oxygen to cells and carries away wastes. Most tissue fluid flows back into the capillaries. Some of it becomes *lymph.

titanium (symbol Ti) A hard white *transition metal *element. Alloys containing titanium are strong, light, and resistant to *corrosion. They are

Tongue

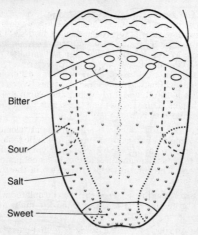

Bitter

Sour

Salt

Sweet

---

used for parts in aircraft and gas *turbines. Titanium dioxide is the *pigment in white paint.

tongue In vertebrates, the organ of taste. The human tongue is covered in *sense organs called taste buds. They are sensitive to either salty, sweet, sour, or bitter tastes.   ✍

tonne A unit of mass equal to 1 000 kilograms.

tooth 1. There can be four different types of tooth in a mammal: *incisor, *canine, *premolar, and *molar.

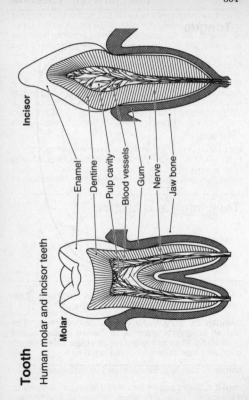

Tooth

Human molar and incisor teeth

Incisor

Molar

Enamel

Dentine

Pulp cavity

Blood vessels

Gum

Nerve

Jaw bone

The crown of a tooth is above the gum and is covered with hard enamel. The root is coated in a thin layer of cement that is attached to fibres growing from the jaw-bone. The pulp cavity contains nerves and blood vessels and is surrounded by bony dentine. 2. see *cog.

torque A measure of the turning effect of a *moment (unit: newton metres, Nm). 1 Nm is equal to a force of 1 N pulling at the end of a lever 1 m long. Torque measures the turning effect of spanners, electric motors, *turbines, and *internal combustion engines.

total internal reflection What happens to a ray of light when an inside surface of a transparent object (e.g. a *prism) acts like a mirror. Total internal reflection results when the ray strikes the surface at

Total internal reflection

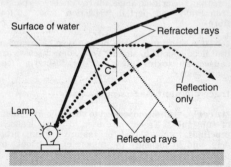

C = critical angle

Transformer

Primary or input coil:
2000 turns

Secondary or
output coil:
100 turns

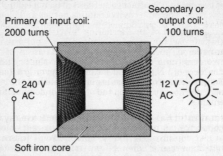

240 V
AC

12 V
AC

Soft iron core

---

an angle greater than the critical angle. At angles
less than the critical angle, the ray undergoes partial
*refraction and partial *reflection. (see *optical
fibre)

toxin A substance made by one type of organism
that is poisonous to another type. Some diseases are
caused by toxins given off by *bacteria. (see
*pathogen)

trace element see *essential element

tracer A *radioisotope used to show the movement
of substances. Tracers are used to detect leaks in
pipelines, to check that *organs are working properly, and to measure the uptake of *nutrients by
plants. The movement of the tracer is followed with
a *Geiger counter.

trachea see *lung

transducer A *device that changes electric signals into mechanical movement, or vice versa. E.g. microphone, *loudspeaker, record-player pick-up.

transformer A device used to change the voltage (and current) of an alternating *electric current. A transformer is made of two coils of wire wound on an iron core. The input (primary) coil acts as an *electromagnet and sets up a changing *magnetic field. *Induction makes a current flow in the output (secondary) coil. The voltage change depends on the relative number of turns on the primary and secondary coils.   ✍

transistor An electronic device made from *semiconductor material and that can be used as a switch or an *amplifier. Transistors usually have three connecting wires. They are used in all types of electronic equipment (e.g. computers, television sets, radios).   ✍

transition metal A *metal *element positioned between groups II and III in the the *periodic table.

Transistor

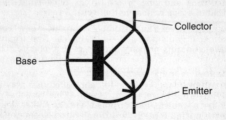

Symbol: npn junction transistor

Transition metals are harder and denser than other metals and have higher melting-points. Examples include *iron, *nickel, and *copper. They have variable *valency and their *compounds are usually coloured.

transition temperature The temperature at which a *physical property of a substance changes. E.g. 96 °C is the transition temperature for rhombic *sulphur to change into monoclinic sulphur.

translucent Describing a substance through which light can pass, but through which objects cannot be seen. A thin sheet of paper is translucent.

transparent Describing a substance through which light can pass and through which objects may also clearly be seen. Water and glass are transparent.

transpiration The loss of water vapour from the leaves of a *plant. Transpiration helps water to flow from the roots, up through the stems and into the leaves. This flow is called the transpiration stream. (see *xylem)

tritium An *isotope of hydrogen, containing two neutrons and one proton in its nucleus. It is a *radioisotope with a *half-life of 12.3 years.

trophic level The position of a *producer or a *consumer organism in a *food chain. The first trophic level contains primary producers (e.g. plants). The second trophic level conains *herbivores. *Carnivores occupy higher trophic levels.

tropism Plants respond to light, water, and gravity by growing in certain directions. This response is called a tropism. In geotropism, gravity makes stems grow up and roots grow down. Hydrotropism causes roots to grow towards water. Phototropism makes stems grow towards the light and roots grow away from it. (see *auxin)

Truth table

OR gate

	Inputs		Output
	A	B	P
Both inputs OFF	0	0	0
One input ON	0	1	1
	1	0	1
Both inputs ON	1	1	1

Output OFF — Output ON

AND gate

	Inputs		Output
	A	B	P
Both inputs OFF	0	0	0
One input ON	0	1	0
	1	0	0
Both inputs ON	1	1	1

Output OFF — Output ON

NOT gate

	Input	Output
	A	P
Input OFF	0	1
Input ON	1	0

Output ON — Output OFF

truth table A diagram that shows how the various input states of a *logic gate affect its output. In a truth table, 0 indicates an input or output that is OFF; 1 indicates ON. Input connections are labelled A, B, etc.; outputs are P, Q, etc. ✍

trypsin An *enzyme that helps with the *digestion of *proteins in food. An inactive form of trypsin is made in the *pancreas. It becomes active after it has entered the *duodenum.

tubule see *kidney

tumour see *cancer

tungsten A grey-white *transition metal element. It has the highest *melting-point (3910 °C) of all the elements. Tungsten is used to make the filaments in electric light bulbs and to make hard *steels for cutting tools. Tungsten carbide is almost as hard as diamond and is used on the edges of some drills and grinding tools.

turbidity A measure of how easy it is to see through a *suspension. Turbidity measurement is used to estimate the cleanliness (solid pollution content) of water samples.

turbine A type of *engine that uses a moving fluid to rotate a shaft; it can also be used as a *machine to compress a gas. A turbine is usually made of angled blades fitted around a shaft. In a turbine engine, water, steam, or burning gases make the shaft rotate by pushing against the blades. (see *hydroelectricity *power-station *turbojet engine)

turbojet engine (jet engine) A type of *internal combustion engine used to move jet aircraft. *Kerosene fuel is burned in air compressed by a *turbine. Hot expanding gases rush from the rear of the engine, producing *jet propulsion.   ✍

turgidity see *turgor

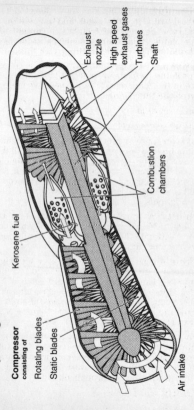

Turbojet engine

Compressor
consisting of

Rotating blades

Static blades

Kerosene fuel

Exhaust nozzle

High speed exhaust gases

Turbines

Shaft

Combustion chambers

Air intake

turgor (turgidity) The state of a plant *cell when *osmosis fills it with water and pushes the *cell membrane against the *cell wall. Turgor in cells keeps plants rigid. Lack of water causes a decrease in turgidity. The result is wilting. (see *plasmolysis)

tympanum see *ear

U

ulna One of the two bones that join the wrist to the elbow in humans and many other vertebrates. (see *skeleton)

ultrasound (ultrasonic sound) Soundlike *waves with *frequencies too high for human ears to detect (i.e. above 20 000 Hz). Uses include ultrasonic cleaners, cutters, and scanners. Medical ultrasonic scanners use reflections from a beam of ultrasound to make images that can show a baby growing in its mother.

ultraviolet (UV) Invisible *electromagnetic radiation present in sunlight. UV radiation is a band of frequencies immediately above the band of visible light in the *electromagnetic spectrum. It causes fair skin to darken and can damage eyes.

umbilical cord In mammals, the tube that connects a growing *embryo to the *placenta. It contains a vein and two arteries. The umbilical cord is cut after birth and shrivels to leave a scar called the navel on the baby.

unicellular Describing an organism made up from a single living *cell. *Bacteria, *protozoa, and some *algae and *fungi are unicellular organisms.

universal indicator see *indicator

Universe All the *matter, *energy, and space that exists. (see *big-bang theory)

unsaturated Describing an *organic compound that contains double or triple bonds. Unsaturated compounds take part in *addition reactions. All *alkenes and *alkynes are unsaturated.

uranium (symbol U) A silvery-white *radioactive metal. Uranium (mostly ^{238}U) contains about 7% of the *isotope ^{235}U. ^{235}U is used as the fuel in *nuclear reactors and in nuclear weapons because slow neutrons cause the nuclei to undergo *nuclear fission. (see *chain reaction)

urea In mammals, a waste substance made in the *liver by breaking down unwanted *protein. Urea is removed from the blood by the *kidneys. Dissolved in water, it flows out of the body as *urine. Pure urea is used as a *fertilizer and to make some *plastics and glues.

ureter In vertebrates, the pipe that carries *urine from the *kidney to the *bladder.

urethra In mammals, the tube that carries *urine from the urinary *bladder to the outside. In male mammals, it also carries *semen from each *vas deferens to the end of the *penis. (see *sexual reproduction)

uric acid A waste substance excreted by birds, insects, and reptiles that live on the land.

urine In animals, the result of *excretion carried out by the *kidneys. The approximate composition of human urine is: water (96%); *urea (2%); *uric acid (0.05%); salts (1.8%).

uterus (womb) In female mammals, the hollow organ where babies grow. The inner wall of the uterus is lined with blood vessels that nourish the growing *embryo through its *placenta. The outer wall is made up from strong muscles that push the offspring out through the *cervix and *vagina during birth. (see *sexual reproduction)

UV Abbreviation for *ultraviolet.

V

vaccinate see *immunize

vacuole A space inside the *cytoplasm of a living *cell. Vacuoles may contain liquids or particles of food.

vacuum A space in which the pressure of gas is less than about 10^{-2} pascals. It is not possible to pump all the gas out of a container and so a perfect vacuum cannot be obtained.

vagina In female mammals, a tube leading from the *uterus to the outside. It receives *semen from the male *penis. Offspring pass through the vagina at birth. (see *sexual reproduction)

valency The number of *electrons (called valency electrons) used by an atom in forming *ions or *covalent bonds in a *compound. The word *valency* is short for *electrovalency* in ionic compounds and *covalency* in covalent compounds. E.g. carbon dioxide O=C=O; each oxygen atom has a valency of 2 and the carbon atom has a valency of 4. The valency of an ion is the same as the number of charges it carries. E.g. sodium Na⁺, valency 1; sulphate SO_4^{2-}, valency 2.

valve 1. A device that controls the flow of a fluid. Water taps, *internal combustion engines, *hearts, and *veins contain valves. 2. see *thermionic valve.

vapour Another name for *gas, especially one which has *evaporated from a liquid.

variable A quantity that can be changed and that affects the results of an experiment. E.g. when measuring the flow of *electric current through a con-

ductor, the *voltage and current are variables. At a
constant temperature, the *resistance of the conductor is fixed and is not a variable.

variation The differences in a *characteristic seen
between individuals in the same *species. Discontinuous variation describes a characteristic which is
either present or not. It is controlled by a small number of *alleles. Blood groups show discontinuous
variation; a person can belong to only one of the four
groups A, B, AB, or O. Continuous variation
describes a characteristic that can be expressed
across a whole range of values. Human height shows
continuous variation and is controlled by a large
number of alleles as well as by environmental factors.

vascular system An arrangement of tubes that
carry liquids around the bodies of living organisms.
In *vertebrates, the liquids are *blood and *lymph. In
plants, the vascular system carries water, salts, and
food substances in the *xylem and *phloem. (see *circulation)

vas deferens In fishes, reptiles, amphibians, birds,
and mammals, a tube that leads *sperms away from
each *testis. (see *sexual reproduction)

vasoconstriction The narrowing of small *blood
vessels when they are squeezed by involuntary
*muscles. Vasoconstriction causes an increase in
*blood pressure.

vasodilation The widening of *blood vessels when
involuntary *muscles that surround them relax.
Vasodilation causes a decrease in *blood pressure.

VDU Abbreviation for *visual display unit.

vector A measurement that shows a quantity or
amount and which also shows a direction. Velocity,

weight, and momentum are vector quantities. (see *scalar)

vegetative reproduction A type of *asexual reproduction used by some plants. A new plant grows from part of the parent plant. This happens, for example, when new bulbs form underground from a parent bulb (e.g. daffodils) or trailing stems grow new roots (e.g. strawberry runners). (see *cutting)

vein 1. A *blood vessel that carries blood towards the *heart. Veins have thinner walls than *arteries. They contain valves which allow the blood to flow in one direction only. All veins except the pulmonary vein contain deoxygenated blood. 2. In the leaf of a plant, bundles of tubes that carry *xylem and *phloem. 3. A hard tube that strengthens the wing of an insect.

velocity A measurement that describes both the *speed of an object and its direction of travel. A car travelling in a straight line at 50 km/hour has a steady velocity of 50 km/hour. A car travelling around a bend at 50 km/hour has a changing velocity because its direction of travel is always changing.

$$\text{Velocity} = \frac{\text{distance moved in a particular direction}}{\text{time taken}}.$$

velocity ratio For any *machine (e.g. see *lever), the distance moved by the *effort compared to the distance moved by the *load.

$$\text{Velocity ratio} = \frac{\text{distance moved by effort}}{\text{distance moved by load}}.$$

vena cava see *heart

venereal disease A disease that is passed from an infected person to another during sexual inter-

ventricle

course. Examples include gonorrhoea, syphilis, and
*AIDS.

ventricle A chamber in the lower part of a *heart.
In mammals, the right ventricle has thick muscular
walls and pumps blood to the *lungs. The left ventricle has very thick walls and pumps blood to the rest
of the body.

vertebra (pl. vertebrae) In vertebrates, the separate
bones that make up the *vertebral column (also
known as the backbone, spine, and spinal column).
There are 33 vertebrae in the human vertebral column. (see *skeleton)

vertebral column (spinal column, spine) In vertebrates, the flexible column of bones that joins to the
skull, the ribs, and the pelvis. It is made up of bones
called *vertebrae which are separated from each
other by *cartilage discs. The *spinal cord runs down
the centre of the vertebral column.   ✍

vertebrate An animal with a backbone (*vertebral
column). Vertebrates include birds, fish, *reptiles,
*mammals, and *amphibians.

villus (pl. villi) A small finger-like organ attached to
the inside wall of the small *intestine. Human villi
contain a network of blood *capillaries. These capillaries absorb substances from digested food and pass
them on to the *liver. *Lymph vessels called lacteals
inside each villus absorb *fatty acids and *glycerol
from the food.   ✍

virus The smallest of all *microbes. Viruses are parasites of plants and animals and some bacteria. They
take control of host cells by injecting *nucleic acids
into them. The host cells manufacture more viruses
and then burst open. Viruses cause diseases such as
the common cold, influenza, polio, and smallpox.

Vertebral column

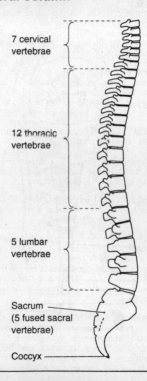

7 cervical
vertebrae

12 thoracic
vertebrae

5 lumbar
vertebrae

Sacrum
(5 fused sacral
vertebrae)

Coccyx

Villus

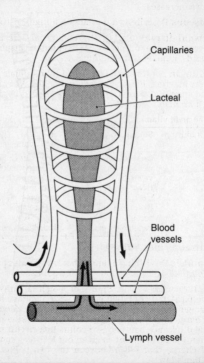

viscosity The 'stickiness' of a liquid, describing its ability to flow. Water has low viscosity and treacle has high viscosity. Viscosity decreases as temperature increases.

viscous Describing a liquid with a high *viscosity.

visual display unit (VDU) A television screen connected to the output of a *computer, displaying words, numbers, or diagrams.

vitamin An *organic compound that animals must take in with their food to remain healthy. Very small amounts of different vitamins are needed by different animals.

The main vitamins required by humans

Vitamin	Source	Result of deficiency
A	carrots, tomatoes	poor sight
B_1	wheatgerm, beans	muscle weakness
B_2	vegetables, milk	poor growth
B_{12}	liver, yeast	anaemia
C	fruit	scurvy
D	liver, fish oil	weak bones
E	cereals	infertility
K	vegetables, eggs	blood does not clot

vitreous humour see *eye

volatile Describing a liquid that *evaporates easily. Volatile liquids have low *boiling-points.

volt (symbol V) The *SI unit of *potential difference (voltage) and *EMF. A battery with a voltage of 1 V provides 1 joule of *energy to every *coulomb of *charge that it pushes around a circuit. If the potential difference between two points in a circuit is 1 V, then 1 joule of energy is given out by every coulomb of charge that flows between the two points. (see *Ohm's law)

Voltameter

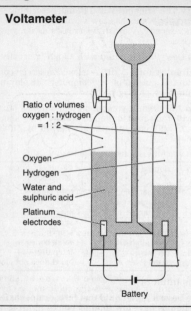

Ratio of volumes
oxygen : hydrogen
= 1 : 2

Oxygen

Hydrogen

Water and
sulphuric acid

Platinum
electrodes

Battery

voltage A measurement of *potential difference, made by connecting a *voltmeter across two points in an electric circuit or across the terminals of a *battery or *generator. (see *volt)

voltameter (Hofmann voltameter) A device used to carry out the *electrolysis of a solution and to collect and measure the volumes of the gases liberated. ✍

Voltmeter

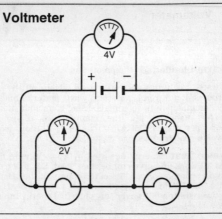

voltmeter An instrument used to measure the *voltage (potential difference) across two points in an electric *circuit. The scale of a voltmeter is marked in *volts.

volume The amount of space taken up by something. The unit of volume is the cubic metre (m³). The volume of a liquid or gas and the volume of the inside of a container are often measured in litres (l), cubic decimetres (dm³), or cubic centimetres (cm³). 1 l = 1 dm³ = 1000 cm³.

vulva The outside opening of the *vagina. In women it is made up from two folds of flesh.

W

warm-blooded see *homoiotherm

waste An unwanted substance given off by a process. E.g. *urea is a natural waste material made by *metabolism in mammals. *Slag is an industrial waste made during the manufacture of iron. Improper disposal of waste causes *pollution. (see *waste heat *radioactive waste)

waste heat *Energy not used by a machine to do useful work and given off by it as *heat. E.g. a petrol engine changes about 15% of the energy in the petrol into *mechanical energy; 85% is lost as waste heat. A *power-station loses over 50% of its fuel energy input as waste heat.

water A colourless liquid that freezes at 0 °C and boils at 100 °C (at *standard pressure). It is an excellent *solvent. Water turns white *anhydrous copper sulphate blue and blue anhydrous cobalt chloride pink.

water cycle The movement of water between the sea, the atmosphere, and the land. Water *vapour is formed in the atmosphere by *transpiration from plants and *evaporation from seas, rivers, lakes, and land. This evaporation is caused by *solar energy. Water vapour *condenses and falls as rain, sleet, or snow.

water of crystallization Water contained in the *lattice of a *crystal. Water of crystallization is incorporated into the crystal as it grows during *crystallization. (see *hydrated)

watt (symbol W) The *SI unit of *power. 1 watt equals 1 *joule of energy expended in one second.

wave The form in which some types of energy (e.g. sound, water waves, light) move from one place to another. Sound-waves are longitudinal waves. Air molecules vibrate (*oscillate) backwards and forwards in the direction the wave is travelling. Water waves are transverse waves. Water oscillates up and down as the wave moves along. The medium (air, water) does not move along with the wave; each particle in the medium oscillates about a fixed point as the wave passes. Light waves (*electromagnetic radiation) are transverse waves. Electric and magnetic *fields oscillate at right angles to the direction the wave is travelling. (Light does not need a medium.) The *velocity, *frequency, and *wavelength of a wave are connected by the equation:

speed = frequency×wavelength.

(see *amplitude *wavefront)

wavefront Lines drawn to represent a progressive *wave. Wavefronts usually show the places of maximum *amplitude of the wave. ✍

Wavefront

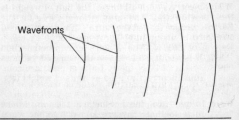

Wavefronts

wavelength For a *wave, the measurement that describes the distance between two adjacent peaks (or between any two equivalent points). ✍

wave mechanics see *quantum theory

wave power Electrical energy generated from the energy of waves in the sea. The waves move floats up and down and drive electric *generators. (see *renewable energy)

wax Natural waxes are solid substances made by some plants and animals for protection against water. E.g. beeswax and the waxy coat on holly leaves. They contain *fatty acids. Mineral waxes for candles and polishes are made from *crude oil. They are *hydrocarbons with between 40 and 50 atoms of carbon per molecule.

weathering The action of wind, temperature changes, chemicals, and water that breaks the surface of *rock into small particles. (see *erosion)

wedge A simple *machine made up of two *inclined planes joined back to back. An axe head is a wedge, used as a cutting machine. A downward force on the axe head creates a strong sideways force that splits the wood (the load).

weight The *gravitational force acting on the *mass of an object. As with all forces, the unit of weight is the *newton (N). The gravitational force on the Earth's surface (called the Earth's gravitational field strength) is about 10 N/kg. A mass of 50 kg has a weight of 500 N. The Moon's gravitational field strength is about 1.6 N/kg, so the same 50 kg mass has a weight of 80 N there. Units of mass are commonly (but incorrectly) used as units of weight (e.g. 'That bag of sugar weighs 1 kg').

weld To melt two pieces of metal at the point where they touch so that they are joined together when

Wavelength

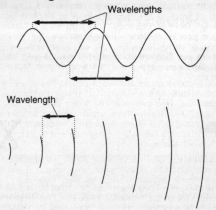

Wavelengths

Wavelength

cool. Gas welding uses heat from *ethyne burning in pure oxygen. Arc welding uses heat from a powerful electric spark.

white dwarf A small star that has used all its fuel and shrunk under its own *gravitational force.

wilting see *turgor

wind power Electrical energy from *generators powered by giant windmills. (see *renewable energy)

wire A long flexible strand of metal (usually copper) coated with plastic *insulation and used to conduct the *electric current around a *circuit.

womb see *uterus

work Work is done when a force moves. The *SI unit of work is the *joule. 1 joule of work is done when a force of 1 newton moves through a distance of 1 metre. Work done = force×distance.

worm An *invertebrate animal with a soft body. The two main types are *flatworms and *annelid worms.

X

xenon (symbol Xe) A gaseous *element belonging to group VIII of the *periodic table (the *noble gases). Xenon is a monatomic gas, composed of separate atoms. It is extracted by *fractional distillation of *liquid air, which contains 0.0009% of the gas. It is used in powerful *discharge lamps.

X-ray A type of *electromagnetic radiation with very high *frequency and short *wavelength. X-rays can penetrate solids and are used to take medical photographs of internal organs and bones. They can cause damage to living cells.

xylem In many types of plant, a *tissue made up from tubes and woody cells. Xylem is part of the *vascular system of a plant and also helps to support stems. It carries water and dissolved *salts from the ends of roots to all other parts of the plant. (see *phloem)

Y

yeast A type of *fungus that exists as single cells. *Saccharomyces* is a type of yeast used in baking and brewing. It uses *anaerobic respiration to break down sugars into carbon dioxide and *ethanol.

yellow spot see *fovea

yield The efficiency of a *chemical reaction in changing *reactants into *products. A 50% yield means that half the reactants are converted into products. The remainder is unchanged or becomes *by-products.

yolk The food stored in an egg and used by the growing *embryo. The largest yolks are contained in bird, reptile, and fish eggs.

Z

zinc A soft grey *transition metal *element. Zinc is made by roasting and *smelting its ore. It is used to make alloys (e.g. *brass) and some types of *battery.

zinc–carbon cell A type of *primary cell made from a carbon and a zinc *electrode in an ammonium chloride *electrolyte. Manganese dioxide is included in the electrolyte as a 'depolarizer' that stops the formation of gas bubbles.

zoology The study of animals.

Zinc carbon cell

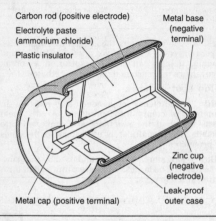

Carbon rod (positive electrode)

Electrolyte paste
(ammonium chloride)

Plastic insulator

Metal base
(negative
terminal)

Zinc cup
(negative
electrode)

Leak-proof
outer case

Metal cap (positive terminal)

---

zygote A single living *cell that results when a female *gamete is fertilized by a male gamete. In plants with *flowers, a zygote results when pollen fertilizes an ovule. The zygote grows to become a seed. In animals, a zygote results when a *sperm fertilizes an *ovum. The zygote grows into a young offspring.

Appendix A: SI units

Physical quantity measured	Name of unit	Symbol
length	metre	m
mass	kilogram	kg
time	second	s
electric current	ampere	A
temperature	kelvin	K
amount of substance	mole	mol
frequency	hertz	Hz
energy	joule	J
force	newton	N
power	watt	W
pressure	pascal	Pa
electric charge	coulomb	C
electric potential difference	volt	V
electric resistance	ohm	Ω
nuclear activity	becquerel	Bq

Appendix B: Abbreviations and symbols

α	alpha
A	ampere
A	mass number (nucleon number)
A_r	relative atomic mass
(aq)	aqueous (dissolved in water)
β	beta
Bq	becquerel
c	speed of light
C	coulomb
°C	degree Celsius
c.c.	cubic centimetre (now cm^3)
cm^3	cubic centimetre
dB	decibel
dm^3	cubic centimetre
e	electron
e^-	electron

E	EMF of a cell
F	farad
F	Faraday constant
°F	degree Fahrenheit
γ	gamma
g	gram
(g)	gas
Hz	hertz
I	electric current
J	joule
K	kelvin
kg	kilogram
kWh	kilowatt-hour
l	litre
(l)	liquid
L	Avogadro number
m	metre
M	mega-
M	molar
mg	milligram
ml	millilitre
M_r	relative molecular mass
mol	mole
n	neutron
N	newton
N_A	Avogadro number
Ω	ohm
p	proton
p$^+$	proton
Pa	pascal
Q	electric charge
ρ	density
s	second
(s)	solid
t	time
$t^{1/2}$	half life
T	temperature
V	volume
V	volt
W	watt
Z	atomic number

Appendix C: Geological time—eras, periods, and epochs

Era	Period	Time range millions of years	Major events
Quaternary	Holocene	0.01–now	temperatures rise; human civilizations
	Pleistocene	2–0.01	Ice Ages; evolution of modern humans
Tertiary (Cainozoic)	Pliocene	5.1–2	temperatures fall; extinction of many mammal species
	Miocene	24.6–5.1	great movements in Earth's crust; Alps and Himalayas start to form
	Oligocene	38–24.6	temperatures fall
	Eocene	54.9–38	temperatures rise; first horses, bats, and whales
	Palaeocene	65–54.9	great increase in mammals
Secondary (Mesozoic)	Cretaceous	144–65	flowering plants; chalk deposits, extinction of dinosaurs
	Jurassic	213–144	largest dinosaurs; birds
	Triassic	248–213	dinosaurs; the first mammals
Primary (Palaeozoic)	Permian	286–248	hot dry climate; reptiles
	Carboniferous	360–286	seed-bearing plants; limestone deposits; origins of coal
	Devonian	408–360	amphibians and forests
	Silurian	438–408	land plants; true fish
	Ordovician	505–438	invertebrates; the first vertebrates (jawless fish)

Era	Period	Time range millions of years	Major events
	Cambrian	590–505	widespread seas; oldest fossils
Pre-Cambrian (Archean)		4500–590	oldest rocks 3800 million years old. origin of Earth 4500 millions years ago